Advanced Indium Arsenide-based HEMT Architectures for Terahertz Applications

Advanced Indium Arsenide-based HEMT Architectures for Terahertz Applications

Edited by
N. Mohankumar

CRC Press
Taylor & Francis Group
Boca Raton London New York

CRC Press is an imprint of the
Taylor & Francis Group, an **informa** business

First edition published 2022
by CRC Press
6000 Broken Sound Parkway NW, Suite 300, Boca Raton, FL 33487-2742

and by CRC Press
2 Park Square, Milton Park, Abingdon, Oxon, OX14 4RN

CRC Press is an imprint of Taylor & Francis Group, LLC

ISBN: 978-0-367-55414-9 (hbk)
ISBN: 978-0-367-55415-6 (pbk)
ISBN: 978-1-003-09342-8 (ebk)

DOI: 10.1201/9781003093428

Typeset in Palatino
by Deanta Global Publishing Services, Chennai, India

Contents

Preface

In this era of high-speed communication, the semiconductor industry is focused on fabricating high-speed devices with reduced short-channel effects and minimal power dissipation. To achieve high-speed performance, III–V compound semiconductors have emerged as a suitable alternative over conventional silicon (Si) channel devices. The semiconductor industry has been dominated by silicon technology for decades with its established complementary metal–oxide–semiconductor (CMOS) process. However, during the scaling of metal–oxide–semiconductor field-effect transistors (MOSFETs) below 22 nm, discrepancies, such as an increase in short-channel effects, affect the performance of the device. To overcome this problem, high electron mobility transistors (HEMTs) arrived as promising candidates because of their reliability when compared to other devices such as silicon nanowires and carbon nanotubes. For applications requiring terahertz (THz) waves, the novel HEMT device with III–V materials, such as InAlAs, InGaAs, and InAs, are used as sources for achieving the terahertz spectrum. In particular, for defense and medical applications, these devices can produce terahertz radiation, loosely defined as having a frequency from 0.1 to 10 THz. This high frequency is made possible by proper device scaling, increasing the concentration of carrier in the channel and achieving low noise from the device. Although scaling of the device gives better performance, it also increases the short-channel effects in the device. To mitigate this effect, a double-gate technology removing the buffer from the single-gate structure HEMT is proposed. This double-gate (DG) structure shows better control over the channel than the conventional single-gate HEMT (SG-HEMT). This DG-HEMT architecture also provides reduced leakage current in the OFF-state current by achieving high current driving capability and high concentration of carriers in the two-dimensional electron gas (2DEG). The five-layered composite channel (InGaAs/InAs/InGaAs/InAs/InGaAs) and optimization of the drain-side recess length of the gate creates a dual quantum well with low lattice defects, increased mobility of carriers, and peak radio frequency (RF) and direct current (DC) performance with reduced low-noise performance.

The entire book elaborates the discussion on the impact of five-layered composite channel and novel device architectures for the growing world of high-frequency electronics. The novel devices are simulated by using TCAD Sentaurus tools, and the results are validated by experimental verification of fabricated SG-HEMT devices. On the whole, this book clearly gives an idea about the influence of dual quantum wells in the DG-HEMT structure and its optimization to achieve frequency near the THz regime with reduced noise.

The motivation for editing this book came about while optimizing the devices for high-frequency applications as my research area and due to the advancement of THz electronics for future detection, imaging, and other fields. This book is intended for senior undergraduate and postgraduate students of electronics engineering, and for researchers and professors working in the area of RF electronic devices. The purpose of this book is to disseminate the concepts and physics of new device architectures with InAs as the channel material to enhance RF performance metrics and realize the feasibility of the devices to work in the THz regime.

The chapters in the book provide a balance between basic concepts and recent technology, and an elaborate literature survey on the devices for high-power, high-breakdown, and high-frequency applications involving new device architectural designs.

Chapter 1 is an introduction to III–V materials and HEMT structures. Chapter 2 discusses the introduction of III–V heterostructure devices for ultralow, high-power, and high-breakdown applications. Chapter 3 presents the III–V heterostructure for high-frequency applications. Chapter 4 provides an overview of THz applications. Chapter 5 discusses the simulation framework of InAs HEMTs. Chapter 6 analyzes the single-gate InAs-based HEMT architecture for THz applications. Chapter 7 realizes the effect of gate scaling on the composite channel in InAs HEMTs. Chapter 8 introduces the double-gate InAs-based HEMT and its advantages over its single-gate counterpart. Chapter 9 covers the influence of dual-channel and drain-side recess length in double-gate InAs HEMTs. Chapter 10 provides insight into the noise analysis in dual quantum well InAs-based DG-HEMTs.

An extensive list of references is provided at the end of every chapter for a deeper understanding and to motivate readers to engage in further research in the RF domain.

I wish to congratulate all the contributors and their peers. Their convictions and efforts are useful in enlightening the minds of the readers. The dedication and efforts of all the contributors led to the successful completion of this book. Special thanks to Gagandeep Singh and Gaba Lakshay and staff members of CRC Press for their patience, guidance, and responsiveness during the book publishing process.

Editor

Dr. N. Mohankumar was born in India in 1978. He earned his BE degree from Bharathiyar University, Tamilnadu, India, in 2000 and ME degree and PhD from Jadavpur University, Kolkata, in 2004 and 2010, respectively. He joined the Nano Device Simulation Laboratory in 2007 and worked as a Senior Research Fellow under CSIR Direct Scheme until September 2009. Later he joined SKP Engineering College as a professor to develop research activities in very large scale integration (VLSI) and nanotechnology. He initiated and formed the Center of Excellence in VLSI and nanotechnology at SKP Engineering College at the cost of 1 crore.

A memorandum of understanding was signed between SKP, the Tokyo Institute of Technology, and the New Jersey Institute of Technology under Dr. Mohankumar's guidance. In 2010 he taught at the Tokyo Institute of Technology as a visiting professor for three months. In 2012 he was at the New Jersey Institute of Technology as a visiting professor for two months. For two weeks in 2013 he was a research professor at National Chiao Tung University (Taiwan). He is currently working as a professor and head of the EECE Department at Gandhi Institute of Technology and Management (GITAM), Bengaluru Campus, in India.

He is a senior member of IEEE and served as secretary of IEEE, EDS Calcutta Chapter, from 2007 to 2010. He served as a chairman of IEEE EDS Madras Chapter from 2010 to 2017. He has published approximately 70 articles in reputed international journals and about 50 international conference proceedings. He received the Career Award for Young Teachers (CAYT) from the All India Council for Technical Education (AICTE), New Delhi, for 2012–2014.

He is a recognized supervisor for research under Anna University and Jadavpur University. Fifteen PhD scholars have completed their studies under his supervision. More than 50 ME students have completed their theses under his guidance. At present, he is guiding three PhD research scholars and two ME students.

His research interests include modeling and simulation study of high electron mobility transistors (HEMTs), optimization of devices for radio frequency applications and characterization of advanced HEMT architecture, terahertz electronics, high-frequency imaging, sensors, and communication.

Contributors

C. Kamalanathan
Electrical, Electronics and
  Communication Engineering
  Department
GITAM School of Technology,
  Bengaluru Campus
Karnataka, India

M. Arun Kumar
Electrical, Electronics and
  Communication Engineering
  Department
GITAM School of Technology,
  Bengaluru Campus
Karnataka, India

R. Saravana Kumar
Electronics and Communication
  Engineering Department
Bannari Amman Institute of
  Technology
Sathyamangalam, Tamilnadu,
  India

V. Mahesh
Electrical, Electronics and
  Communication Engineering
  Department
GITAM School of Technology,
  Bengaluru Campus
Karnataka, India

Sanhita Manna
Electrical, Electronics and
  Communication Engineering
  Department
GITAM School of Technology,
  Bengaluru Campus
Karnataka, India

Girish Shankar Mishra
Electrical, Electronics and
  Communication Engineering
  Department
GITAM School of Technology,
  Bengaluru Campus
Karnataka, India

N. Mohankumar
Electrical, Electronics and
  Communication Engineering
  Department
GITAM School of Technology,
  Bengaluru Campus
Karnataka, India

T. Nagarjuna
Electrical, Electronics and
  Communication Engineering
  Department
GITAM School of Technology,
  Bengaluru Campus
Karnataka, India

R. Poornachandran
Department of Electronics and
    Communication Engineering
V.S.B. Engineering College
Karur, Tamilnadu, India

D. Godwin Raj
Electronics and Communication
    Department
Amal Jyothi College of Engineering
Kanjirapally, Kottayam, Kerala

Y. Vamsidhar
Computer Science Engineering
    Department
GITAM School of Technology,
    Bengaluru Campus
Karnataka, India

1

Introduction to III–V Materials and HEMT Structure

Sanhita Manna

CONTENTS

1.1 Introduction

For more than a century, silicon (Si) material has played a crucial role by shouldering the responsibility in the semiconductor industry. Low-power metal–oxide–semiconductor field-effect transistors (MOSFETs) fabricated with silicon semiconductors are widely used in the electronics sector. Other devices based on Si include bipolar junction transistors (BJTs), vertical FETs, insulated gate bipolar transistors (IGBTs), and laterally diffused MOSFETs (LDMOS). These devices are exclusively used in low-power switching and

DOI: 10.1201/9781003093428-1

low-noise amplifiers due to their high speed of operation. Despite dominating the market, their use has reached the end of the road, because of performance limitations due to scaling and Si material constraints. But the demand from the microelectronics and communications industries is increasing rapidly in terms of scaling and performance. Researchers are looking for suitable materials to overcome these constraints, with narrow bandgap semiconductors such as indium phosphide (InP) and indium arsenide (InAs) paving the way for future electronics.

Over the last decade, monolithic microwave integrated circuits (MMICs) have been used for microwave and radio frequency (RF) applications. III–V compound semiconductor-based devices such as gallium arsenide (GaAs), metal–semiconductor field-effect transistors (MESFETs), GaAs high electron mobility transistors (HEMTs), and InP heterojunction bipolar transistors (HBTs) are preferred due to their outstanding carrier transport properties. They are widely used in satellite receivers, low-power amplifiers, and the defense industry.

High-technology insights and scaling have increased the need to fabricate devices with high-speed operation, low-power dissipation, and reduced short-channel effects. Moreover, silicon-based devices have already reached their limits for further downscaling. So researchers need to look for an alternative material to meet the increasing demand of industries. This has paved the way for new III–V compound semiconductor-based devices with narrow bandgap energy and high mobility with low noise at higher frequencies.

1.2 Semiconductors Based on III–V Materials

Advancements in the fabrication of nanodevices created an opportunity for new materials to replace silicon in the semiconductor industry. III–V compound semiconductor materials emerged as proper candidates owing to their peak electron mobility, optimal gate control, and high-frequency applications (Dae-Hyun Kim 2008). One elemental semiconductor from the III column of the periodic table and another from the V column are combined to form a III–V compound semiconductor. The III–V materials and their compounds are shown in the periodic table in Figure 1.1.

To fabricate III–V compound semiconductors, three types of combinations are possible: binary (InAs), ternary (InGaAs), and quaternary (InGaAsP). These types of compound semiconductors are utilized to enhance the performance of a device. The electron mobility of indium gallium arsenide (InGaAs) and indium arsenide (InAs) is ten times higher than the conventional Si (Del Alamo 2011). The comparison of III–V

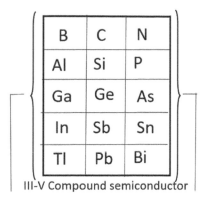

FIGURE 1.1
The III and V group elements in the periodic table is an alloy of semiconductors; different material compounds can be explored with a wide range of energy bandgap.

TABLE 1.1

Comparison of Material Properties of Si and Ge along with III–V Compound Semiconductors

Material Property	Si	Ge	GaAs	InAs	$In_{0.53}Ga_{0.47}As$
Electron effective mass, m_e	0.19	0.08	0.063	0.023	0.041
Electron mobility, M_e (Cm^2/V S)	1450	3900	9200	21000	12000
Hole mobility, M_h (Cm^2/V S)	370	1800	400	450	300
Bandgap, E_g (eV)	1.12	0.66	1.42	0.35	0.74
Relative permittivity, ε_r	11.7	16.2	12.9	15.2	13.9
Lattice constant, a_0 (Å)	5.43	5.66	5.65	6.06	5.87
Thermal expansion coefficient (10^{-60} c)	2.6	5.9	5.73	4.52	5.66

compound semiconductors, along with silicon and germanium, is shown in Table 1.1.

For digital applications, the switching speed of the device must be high. High ON-state current (I_{ON}) and low OFF-state current (I_{OFF}) are the two crucial factors determining the device switching speed. The injection velocity of electrons and the sheet carrier density help in obtaining high I_{ON} from the device. A comparison of injection velocity in nanoscale III–V semiconductors and silicon with their respective gate length is illustrated in Figure 1.2. The figure shows that III–V semiconductors possess higher injection velocity than their Si counterparts.

Due to their outstanding carrier transport properties, III–V compound semiconductors are preferred to fabricate various devices like MOSFETs, metal–insulator–semiconductor field-effect transistors (MISFETs), HBTs, and HEMTs.

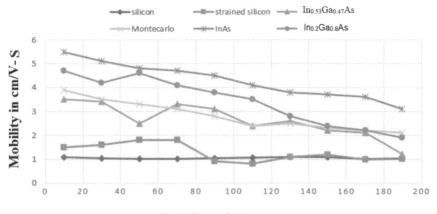

FIGURE 1.2
Apparent and intrinsic channel mobility versus gate length (L) for silicon-, strained silicon-, Monte Carlo-, InGaAs-, InAs-, and InGaAs-based devices fabricated for low-power applications.

1.3 Heterojunction Structure

The HEMT structure is formed by combining epitaxially grown layers with the difference in bandgap energy and composition. The variance in bandgap energy (E_g) and the lattice constant of the semiconductor material used are the two critical parameters that define the heterostructure. The lattice constant and bandgap energy of various III–V compound semiconductors are shown in Table 1.2.

1.4 Discontinuity in Bands

The most influential parameter in the aspect of heterojunction is the occurrence of discontinuity in energy bands. This energy band discontinuity helps to modify the carrier transport in heterostructure devices. The energy band diagrams of two different semiconductors are given in Figure 1.3. Here, E_c and E_v are the energies of the conduction band and valence band; and E_{g1} and E_{g2} are the bandgap energies of the two semiconductor materials used. χ is the electron affinity, and E_F denotes the Fermi level. ΔE_c is the bandgap discontinuity in the conduction band, whereas ΔE_v is the bandgap discontinuity in the valence band (Wu et al. 2008).

TABLE 1.2

Lattice Constant And Energy Bandgap of Various III–V Compound Semiconductors

Alloy	Lattice Constant, a_0 (Å)	Bandgap Energy, E_g (eV)
GaAs	5.653	1.42
AlAs	5.660	2.16
InAs	6.058	0.37
InP	5.869	1.35
$In_{0.53}Ga_{0.47}As$	5.869	0.76
$In_{0.52}Al_{0.48}As$	5.869	1.48
$In_{0.7}Ga_{0.3}As$	5.869	0.58

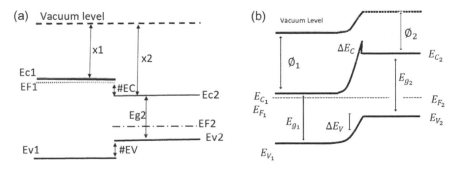

FIGURE 1.3
Energy band diagram of wide bandgap and narrow bandgap semiconductor (a) before and (b) after contact.

For the heterostructure, the bandgap energy model was first devised by Anderson. In this model, the difference in electron affinities was considered to match the value of ΔE_c, which is given by Equations 1.1 and 1.2:

$$\Delta E_c = \chi_1 - \chi_2 \tag{1.1}$$

$$\Delta E_V = \left(E_{g2} - E_{g1}\right) - \left(\chi_1 - \chi_2\right) \tag{1.2}$$

By further reducing the preceding equation, we get Equations 1.3 and 1.4:

$$\Delta E_g = E_{g1} - E_{g2} \tag{1.3}$$

$$\Delta E_g = \Delta E_c + DE_v \tag{1.4}$$

The bandgap engineering in the semiconductor is an essential feature because it produces a large number of discontinuities at the junction interface. The ΔE_c value of InGaAs/InAlAs (>0.5 eV) is high when compared to the other materials, such as AlGaAs/GaAs (~0.3 eV). Because of this exclusive advantage, InGaAs, InAlAs, and InAs materials emerged as the prominent alternatives for upcoming high-speed device applications showing better resilience over the control of charge carriers at the junction interface.

1.5 Triangular Quantum Well and 2DEG

A triangular quantum well (QW) is formed in a heterostructure device by sandwiching a low bandgap semiconductor material (InAs) with two high bandgap semiconductor materials (InGaAs). Due to this, discontinuity in the energy band occurs at the edge of the valence band and conduction band, creating a quantum well. The carriers get accumulated in the well as displayed in Figure 1.4. The layers with high bandgap energy should be heavily doped to supply the carriers to the quantum well. This phenomenon leads to the formation of two-dimensional electron gas (2DEG).

The electrons in the triangular quantum well move parallel to the interface; hence, it is quasi-two-dimensional (Waldron et al. 2010). By choosing the proper material and its combination, the carrier density in the 2DEG quantum well can be controlled.

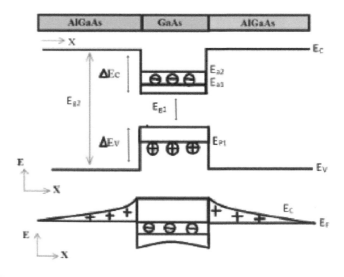

FIGURE 1.4
The formation quantum well with (a) structure, (b) energy band diagram, and (c) conduction band energy.

1.6 Contacts in Semiconductors

In every semiconductor device, metal–semiconductor contacts are always available. Depending upon the nature of the interface, the contacts for the metal–semiconductor interface are divided into two types: ohmic contact and Schottky contact.

1.6.1 Ohmic Contact

If there is no potential barrier between the metal and semiconductor interface, it is called an ohmic contact. This is a non-rectifying contact and does not control the flow of current. Thus, it makes the current flow equally in both the forward and backward direction with good linear I–V characteristics.

1) The work function of the metal (ϕ_m) must be nearer or lower than that of the electron affinity (χ) in n-type semiconductors. So, for n-type, the ohmic contact can be formed by keeping the metal work function lower than that of the semiconductor, i.e., $\phi_s > \phi_m$. This is visible in Figure 1.5a.

2) The work function of the metal (ϕ_m) must be nearer or higher than that of the electron affinity (χ) in p-type semiconductors. So, for the p-type, the ohmic contact can be obtained by maintaining the metal work function higher than that of the semiconductor, i.e., $\phi_s < \phi_m$. Thus, it makes the fabrication process of p-type ohmic contact a tedious one when compared to the n-type ohmic contact. This is visible in Figure 1.5b.

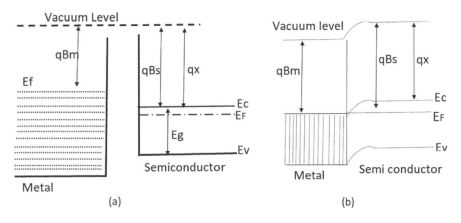

FIGURE 1.5
Bandgap energy diagram of ohmic contact (a) before and (b) after contact.

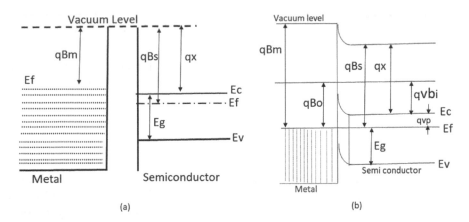

FIGURE 1.6
Bandgap energy diagram of Schottky contact (a) before and (b) after contact.

1.6.2 Schottky Contact

If there is a potential barrier between the metal and semiconductor contact, it is defined as the Schottky contact. This contact is a rectifying one, and hence it permits the flow of current. Because of the influence of the barrier, the flow of current will be in one direction, i.e., in the opposite direction. The semiconductor work function should be lower than that of the metalwork function, i.e., $\phi_s < \phi_m$, to establish a Schottky contact.

The semiconductor/interface before and after the formation of Schottky contact typical for the n-type semiconductor is shown in Figure 1.6. During the Schottky contact, lower energy states of metal will receive electrons from the semiconductors conduction band until a constant value of the Fermi level is achieved at equilibrium condition.

A thick layer of the depletion region is formed due to the transfer of electrons leaving behind the ionized donor's positive charge in the semiconductor device, as depicted in Figure 1.6. The potential barrier (ϕ_b) is due to the band bending at the equilibrium. The built-in potential at the interface can restrict the diffusion of electrons into the metal from the semiconductor. By deciphering the Poisson and Schrödinger equations, the valence band and conduction band can be specified.

1.7 High Electron Mobility Transistor (HEMT)

By joining two different semiconductor materials with two different bandgaps, band bending is observed at the interface due to discontinuity, leading to a quantum well in the channel. The electrons that get accumulated in the quantum well can move in a two-dimensional plane, hence it is called 2DEG. Significantly, the sheet carrier density is improved with reduced scattering, thereby increasing the drive current of the device.

1.7.1 Epitaxial Layers in the HEMT

A substrate with semi-insulating properties is generally used in the HEMT device structure. The epitaxial layers, such as the buffer, channel, barrier, and cap, are grown. These layers have different bandgap energies, doping concentration, and layer thickness in terms of their material properties, as depicted in Figure 1.7.

1.7.2 Cap Layer

The cap layer is used to form contact with the source and drain metal in the structure. The cap layer should be heavily doped to create low resistance at the source and gate metal contacts ($>10^{18}$ cm^3). The overall thickness of the cap layer should be between 10 Å and 100 Å to obtain better performance (Hamaizia et al. 2011). This reduces the contact resistance leading to an increment in the velocity of electrons, transconductance (g_m), and cutoff frequency (f_T).

1.7.3 Barrier Layer

The wide bandgap material is used as the barrier layer, and it is formed below the cap layer. This layer should be doped uniformly by using silicon to diffuse the carriers directly into the channel of the device, increasing the conductivity of the channel. The barrier thickness plays a crucial role

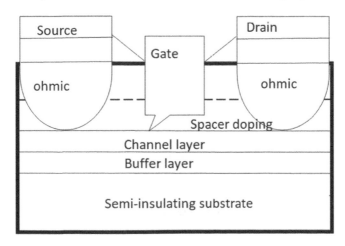

FIGURE 1.7
The metal–semiconductor field-effect transistor (MESFET) consists of a conducting channel positioned between a source and drain contact, and Schottky metal gates control the carrier flow.

in determining the distance between the channel and gate (Mondal et al. 2018). The small gap between the metal and the channel is made preferable by incorporating a thin barrier layer in the device structure.

1.7.4 δ-Doping Layer

Generally, the barrier is uniformly doped, and this can be replaced by introducing a thin δ-doped layer below the barrier. This plays a critical role by reducing parallel conduction in the barrier, thereby improving the breakdown voltage and sheet carrier density of the device. Thus, the carrier concentration in the channel is increased further (Sexl et al. 1997). The electrons from the barrier layer tunnel through the potential barrier and then get trapped in the quantum well. The 2DEG formed at the bottom of the quantum well helps to achieve high mobility in the device. The spacer layer separates the Coulomb scattering between the standard ionized atoms and electrons to further improve carrier mobility. The relation between the drain current and the δ-doping concentration is illustrated in Figure 1.8. The figure shows that the drain current is increased for higher δ-doping concentration.

1.7.5 Spacer Layer

The undoped material is used for the spacer layer between the barrier and channel to differentiate the negatively charged 2DEG from the ionized

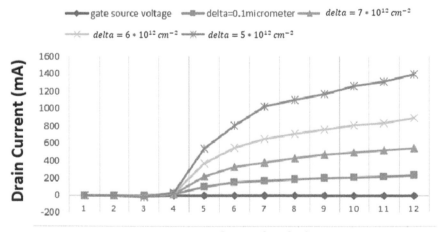

FIGURE 1.8
Experimental drain current versus gate voltage characteristics as a function of substrate bias at room temperature with steps of delta values 5×10^{12} cm^{-2}, 6×10^{12} cm^{-2}, and 0.1 micrometers are plotted for various values of voltages.

dopant atoms. For power devices and low-noise applications, a thin spacer layer is preferable because of its capability to achieve reduced parasitic resistance, peak g_m, and high current density. The noise characteristics of HEMTs at cryogenic temperatures mainly depend upon the thickness of the spacer layer due to its peak electron mobility and velocity (Chang et al. 2007).

1.7.6 Channel Layer

The undoped narrow bandgap material is introduced for the channel layer. The high channel conductivity is mainly due to considerable carrier accumulation in the quantum well. A higher concentration of indium in the channel will significantly improve the electron carrier confinement and transport properties of the HEMT device due to its high electron mobility (Ajayan and Nirmal 2017).

1.7.7 Buffer Layer

The buffer layer aims to mitigate the various defects from the substrate and confine the carriers in the channel (Joo et al. 2008). An undoped wide bandgap material is used to form the buffer layer producing an energy barrier in the conduction band. It thus minimizes the intrusion of electrons into the buffer or substrate. The device growth defects are reduced by using a thick buffer layer, and it also collects any impurities from the substrate that may affect the performance of the 2DEG channel.

1.8 Summary

The transport properties of III–V compound semiconductors are utilized to fabricate MOSFETs, MISFETs, HBTs, and HEMTs. The device heterostructure is defined by two critical parameters: the variance in bandgap energy (E_g) and the lattice constant of the semiconductor material. Significantly, the band discontinuity helps to adjust the carrier transport in the HEMT device. It also develops the triangular QW, which uses the Schottky and ohmic contacts for the metal–semiconductor junction. The ohmic contact is classified as p-type and n-type. In a Schottky contact, lower energy states of metal will receive electrons from the semiconductor's conduction band until a steady Fermi level matching is executed at equilibrium circumstances. The substrate with semi-insulating properties acts as a base for epitaxial growth of the buffer, channel, barrier, and cap layer. The cap layer reduces the source and drain spacing, decreasing the contact resistance leading to an increment in the velocity of electrons, transconductance (g_m), and cutoff frequency (f_T). Moreover, it is observed that the drain current is increased with a δ-doping concentration in the respective layer.

References

J. Ajayan, D. Nirmal, 2017, "22 nm $In_{0.75}Ga_{0.25}As$ Channel-Based HEMTs on InP/GaAs Substrates for Future THz Applications," *Journal of Semiconductors*, 38, 44001.

C. Chang, H. Hsu, E.Y. Chang, C. Kuo, S. Datta, M. Radosavljevic, Y. Miyamoto, G. Huang, 2007, "Investigation of Impact Ionization in InAs-Channel HEMT for High-Speed and Low-Power Applications," *IEEE Electron Device Letters*, 28, 856–858.

Jesus A. Del Alamo, 2011, "Nanometre-Scale Electronics with III-V Compound Semiconductors," *Nature*, 479, 317–323.

Z. Hamaizia, N. Sengouga, M. Missous, M.C.E. Yagoub, 2011, "A 0.4dB Noise Figure Wideband Low-Noise Amplifier Using a Novel InGaAs/InAlAs/InP Device," *Materials Science in Semiconductor Processing*, 14(2), 89–93.

K.S. Joo, S.H. Chun, J.Y. Lim, J.D. Song, J.Y. Chang, 2008, "Metamorphic Growth of InAlAs/InGaAs MQW and InAs HEMT Structures on GaAs," *Physica E: Low-Dimensional Systems and Nanostructures*, 40, 2874–2878.

D.H. Kim, J.A. Alamo, 2008, "Lateral and Vertical Scaling of $In_{0.7}Ga_{0.3}As$ HEMTs for Post Si-CMOS Logic Applications," *IEEE Transaction of Electronic Devices*, 55, 2546–2553.

S. Mondal, S. Paul, A. Sarkar, 2018, "Investigation of the Effect of Barrier Layer Engineering on DC and RF Performance of Gate-Recessed AlGaN/GaN HEMT BT," in *Methodologies and Application Issues of Contemporary Computing Framework*, edited by J.K. Mandal, S. Mukhopadhyay, P. Dutta, K. Dasgupta, 177–184. Singapore: Springer.

M. Sexl, G. Böhm, D. Xu, H. Heiß, S. Kraus, G. Tränkle, G. Weimann, 1997, "MBE Growth of Double-Sided Doped InAlAsInGaAs HEMTs with an InAs Layer Inserted in the Channel," *Journal of Crystal Growth*, 175–176, 915–918.

N. Waldron, D. Kim, J.A. Del Alamo, 2010, "A Self-Aligned InGaAs HEMT Architecture for Logic Applications," *IEEE Transactions on Electron Devices*, 57, 297–304.

C.Y. Wu, H.T. Hsu, C.I. Kuo, E.Y. Chang, Y.L. Chen, 2008, "Evaluation of RF and Logic Performance for 40 nm InAs/InGaAs Composite Channel HEMTs for High-Speed and Low-Voltage Applications," *Proceedings in Asia Pacific Microwave Conference APMC*, 75–78.

2

III–V Heterostructure Devices for Ultralow-Power, High-Power, and High-Breakdown Applications

D. Godwinraj

CONTENTS

2.1 Introduction

2.1.1 Indium Antimonide HEMTs for Low-Power Applications

The unit cell crystal structure of indium antimonide (InSb) is zinc blende, consisting of two face-centered cubic lattices displaced along a diagonal from each other. The two sublattices are made up of indium (group III) and antimony (group V) (Mamaluy and Gao 2015).

DOI: 10.1201/9781003093428-2

TABLE 2.1

Properties of High Mobility III–V Semiconductor Materials
at Room Temperature

Property	InSb	InAs	GaAs	Si
Electron g-factor	−0.51	−15	−0.44	1.99
Electron mobility (μ_n, cm^{-2}/V-S)	77,000	40,000	9200	1600
Effective mass (m*/m$_o$)	0.014	0.023	0.065	0.19
Energy gap (eV)	0.175	0.356	1.43	1.12
Dresselhaus coefficient (eV Å)	760	27.2	27.6	144

InSb shows incredible potential for ultrafast, low-power consumption as it has high electron mobility offering high speed better than any known semi-conductor (see comparison in Table 2.1). The fabrication and characteriza-tion of an InSb channel quantum well field-effect transistor (FET) utilize a semiprotecting gallium arsenide (GaAs) substrate, a tunable barrier layer of $Al_yIn_{1-y}Sb$, a compressively strained InSb layer between $Al_xIn_{1-x}Sb$ layer, and a Schottky metal gate.

2.1.2 Failure Mechanism in InSb-Based Heterostructures

InSb gained attention due to its direct bandgap ~0.17 eV, lowest effective mass (0.013 m$_e$), and high room-temperature electron mobility (77000 cm^2/V-s). Hence, it is suitable for various device applications, including magneto-resistive sensors devices and infrared devices. GaAs substrate offers better insulating behavior and is cheaper than InSb wafers. Metal organic chemi-cal vapor deposition (MOCVD) and molecular beam epitaxy (MBE) are the fabrication techniques for growing InSb/GaAs heterostructures (M. Yano et al. 1979). Grown GaAs substrate and InSb film exhibit a high mismatch of 14.6%. The lattice mismatch between GaAs substrate and InSb film creates a defective failure mechanism for many applications. The electron mobility is much degraded because of the interface roughness between the buffer and substrate region, causing predominant scattering.

2.1.3 Effects of Threading Dislocation Degradation

High dislocation density and strain–relaxation-induced misfit defects occur between InSb film and GaAs substrate. Threading dislocation defects are one of the main defect factors that affect mobility beyond 80,000 cm^2/V-s. Maximum dislocation density is reportedly as high as 10^{11}/cm^2 at these inter-faces (X. Zhang et al. 1990). These dislocations increase the coulomb poten-tial scattering and deformable potential scattering. These two scatterings directly affect the free carriers through depletion of potential areas or lattice dilation associated with these regions.

Inversion symmetry in III–V compound semiconductors related to the dislocation reduces the electron mobility of the device. Earlier studies reported the drastic degradation of electron mobility in InSb film with 1 mm thickness at high temperatures (C. J. Kiely et al. 1989). The doping slab at the InSb/GaAs interface reduces the mobility twice for the same carrier doping than a slab inserted 0.75 mm away from the interface (X. Zhang et al. 1990). These effects are directly related to threading dislocation density due to lattice defects. However, very few studies have been reported considering the effects of dislocation scattering (M. J. Yang et al. 2000).

Simultaneously, dislocation effects and roughness properties of film/substrate devices are not well understood. Better understandings of these effects are essential for electronic device applications. We can examine these effects through a literature survey of highly mismatched InSb films grown by MOCVD on GaAs substrates (M. J. Yang et al. 2000).

From an experimental review, we can observe that threading dislocation density decreases with increasing film thickness (M. J. Yang et al. 2000). These results are consistent with a significant increase in electron mobility. They also suggest that the threading dislocation density is the dominant factor of electron mobility degradation at room temperature. High lattice mismatch in these devices should be avoided to minimize the threading dislocation and lattice degradation. Alternative material or incorporation of alternative material in InSb is highly desirable to reduce the lattice-mismatched effect.

2.1.4 Remedies to Address Defect Issues in InSb HEMTs

The electron mobility degrades drastically from low to room temperature. From the aforementioned analysis, the possible solution to overcome threading dislocation defects is by including As with an InSb binary compound. To form a heterointerface with InSb, we need a wide bandgap material with a better lattice match with InSb. AlSb barrier is a closer wide bandgap material, which can be a better match with the InSb channel. Figure 2.1 shows that InAs has a much better lattice match with AlSb. So the mole fraction of InAs to InSb will be a better solution for a better heterointerface.

Variation of the lattice constant from 6.2 Å to a lower value will improve the device carrier band offset. Impact ionization is significant for electron confinement in the narrow bandgap semiconductor device. Incorporation of As to Insb increases the valence band offset for the InAsSb resulting in a perfect type I band alignment. For digitally grown InAsSb quantum wells, the band alignment is type I when the Sb atom fraction is 15% (P. B. Klein and Binari 2003). The lattice mismatch between the buffer and the channel is degraded by introducing AlInSb in the barrier and InAsSb in the channel (P. B. Klein and Binari 2003).

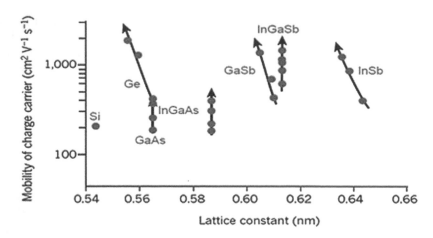

FIGURE 2.1
Variation of III–V material lattice constant (Å) with mobility of charge carriers (cm²v⁻¹s⁻¹).

2.2 III–V Heterostructure Devices for Ultrahigh-Power Applications

2.2.1 Wide Bandgap HEMTs for High-Power, High-Temperature Electronics

Power management is a fundamental requirement to manage a device that conducts electricity. As such, advancements in power devices enable wide bandgap (WBG) materials in power electronics. Leveraging a history that began with light-emitting diodes (LEDs) and then expanded into radio frequency (RF) devices with SAW filters, WBG semiconductors were introduced in the power electronics world in 1992 with the demonstration of the first 400 V silicon carbide (SiC) Schottky diode. Since then, the WBG power electronics portfolio has expanded to include 1200 V SiC Schottky diodes and rectifiers, junction–gate field-effect transistors (JFETs), metal–oxide–semiconductor field-effect transistors (MOSFETs), bipolar junction transistors (BJTs), and thyristors from several manufacturers, including Cree and ST Microelectronics.

SiC is the most maturely developed WBG material, as evidenced by the number of SiC power devices available today from various manufacturers, including Cree, GeneSiC, Infineon, Panasonic, ROHM, ST Microelectronics, Semelab/TT Electronics, and Central Semiconductor. The advantages of SiC over silicon for power devices include lower losses and higher efficiency, higher switching frequencies to trim down passive components for more compact designs, and higher breakdown voltages. SiC enables higher operating speeds and smaller-sized magnetics in power electronic designs. SiC also exhibits significantly higher (3x) thermal conductivity than silicon, with temperature having little influence on its switching performance and thermal characteristics such as ON-resistance. The low loss, high-efficiency operation of SiC devices in temperatures beyond 150°C, the maximum operating temperature of silicon, and reduction in cooling/thermal management requirements (elimination of fans and heat sinks, for example) lower system cost, and smaller form factors are also advantages. The high thermal conductivity also lends to the robustness of SiC devices. High thermal conductivity combined with the homogeneous substrate and epitaxy layers (SiC devices are built on a SiC substrate) allows for vertical power devices that can distribute heat effectively across the die and withstand high current surges and high transient voltages. These properties make SiC ideal for high-power (>1200 V, >100 kW), high-temperature (200°C–400°C) applications, but also suitable for less stringent usage as well. Renewable energy generation (solar inverters and wind turbines), geothermal (down-hole drilling), automotive (hybrid/electric vehicles), transportation (aircraft, ships, and rail traction), military systems, space programs, industrial motor drives, uninterruptable power supplies, and power factor correction (PFC) boost stages in offline power supplies are all suitable applications of SiC power devices.

SiC-based power semiconductors accounted for about US$200 million in sales in 2013 but are expected to take off sharply in the next few years and, by some projections, will near US$1.8 billion by 2022. With GaN power devices entering the market, sales are projected to top US$1 billion in that same period. Despite these exponential growth expectations for SiC and GaN power devices, given that the global power semiconductor market is estimated to be worth around US$65 billion in 2020 (per IHS/ IMS Research), it is clear that opportunities for silicon will continue for years to come. It is especially true in the low power, low voltage market where a low bandgap is desirable. Development in new materials for the low power, low voltage market is still in infancy. For example, graphene, a zero-bandgap material, has generated a lot of excitement due to its unique properties. Graphene has a tunable bandgap, excellent conductivity, durability, lightweight, and was recently isolated in 2004. Interestingly, heating SiC to high temperatures (>1100°C) under low pressures (~10^{-6} torr) reduces it to graphene.

WBG materials have a long history of LEDs. LED activity utilizing SiC was initially exhibited in 1907, and the first eras of commercial LEDs that were accessible from the 1960s through the 1980s were because of SiC. In the mid-1990s, basic advancements in GaN demonstrated that it could deliver 10 to 100 times brighter outflows than SiC. The achievement of first entry high-splendor blue LEDs came from the strong beginning of the state lighting industry because of white lighting created by blue LEDs covered in phosphor. By 2020, IHS/IMS Research states that LEDs will be found in four out of five attachments in created locales, with radiant globules covering only 2% and compact fluorescent lamps (CFLs) compensating for any shortfall.

Insulated gate bipolar transistors (IGBTs) are dominated by GaN and SiC-based microwave and force gadget execution. They are more efficient than silicon-based IGBT gadgets. For power speakers, complex power intensifiers, telecom applications, vehicles, engine applications, and mechanical high-power hardware, GaN-based devices dominate the industry. Both typically ON and OFF GaN-based HEMTs have replaced other semiconductor devices in the power electronics industry, as depicted in Figure 2.2.

GaN-based HEMTs have emerged as a potential candidate for future power electronics due to their unique and attractive properties such as high two-dimensional electron gas (2DEG) sheet carrier density (n_s), high-switching rate, high 2DEG portability, low ON-resistance and high-temperature dependability, high bandgap, and high recurrence of operation. GaN-based HEMTs have a higher breakdown voltage ten times greater than Si, and higher bandgap, higher versatility, and higher transport properties.

In 2011, Cree dispatched its Z-FET™ line of SiC MOSFETs, giving record efficiencies while enhancing unwavering quality in speed switching applications. ST Microelectronics's STPSC groups of SiC diodes are accessible at 600 V, 650 V, and 1200 V. STPSC6H12 is an elite 1200 V SiC Schottky rectifier that is particularly intended to be used as a part of photovoltaic inverters. It helps to build the inverter yield by up to 2% on account of its capacity to work at high recurrence with low switching hazards and ultrafast switching at all temperatures. The hazards created by SiC diodes are 70% lower contrasted with bipolar diodes.

WBG materials additionally emit light, and this optical property has influenced the LED manufacturing industry. GaN-based LEDs give an energy

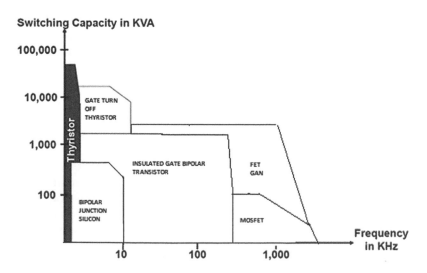

FIGURE 2.2
The power comparison of GaN devices and the relationship between switching capacity and frequency.

efficient, tough, long-life utilization. GaN is also utilized as a part of laser diodes, with the most unmistakable usage today is a Blue-beam player.

GaN devices grown over a SiC or silicon substrate was introduced because of the restrictive cost of utilizing a homogeneous GaN substrate. As expenses go down, GaN power gadgets will be feasible for future power devices.

Improving the performance of wide bandgap semiconductor materials and devices has been progressing for a long time. The material properties of wide bandgap materials are especially attractive to device designers because they guarantee higher performance benefits over their silicon-based counterparts. These include high-breakdown voltage, high 2DEG carrier density, and high-temperature operation, making these devices profoundly fascinating for future high-power electronic applications. GaN and SiC-4H are two such wide-bandgap promising devices that can be used for switching and RF applications. The low ON-resistance (R_{ON}) along with high-breakdown voltage comparison is shown in Figure 2.3.

For higher voltages and lower leakage current operations, both GaN and SiC are better than Si due to the high operating electric field. GaN-based HEMT devices are the best fit for high-speed applications due to their higher 2DEG mobility than SiC and silicon materials. Whereas SiC has higher thermal conductivity with efficient heat dissipation capability, GaN and silicon have lower thermal conductivity in contrast to SiC, which implies better high-power operation for SiC-based devices than GaN- and silicon-based devices. The blend of high bandgap, high electric field, and higher thermal conductivity collectively provide SiC semiconductor material an advantage

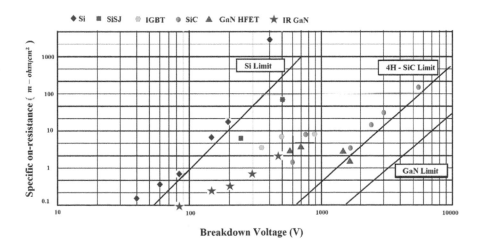

FIGURE 2.3
The characteristics between breakdown voltage and specific ON-resistance for different HEMT devices. Breakdown voltage increases proportionate changes in the specific resistance in Si, SiC, IR GaN, and SiSJ similarly increase in specific resistance and saturation in IGBTs and GaN HFETs.

when ultrahigh power is a crucial requirement. Heat dissipation of GaN is a technological challenge for device designers. However, the versatility and polarization charge density of GaN provide extra advantages for high-power applications. This breakdown voltage can be further increased using a gate field plate and source to drain field plate technologies.

2.3 III–V GaN-Based Compound Semiconductors

The III–V nitrides (AlGaN, InAlN, InN, GaN, and their reliable solutions) are unique with their high sheet carrier concentration at heterojunction interfaces. They possess a wide bandgap, high saturation velocity, electron mobility, high-breakdown field, and low thermal impedance when grown over SiC substrates (Chih-Hsien Cheng et al. 1978). They also possess the merits of chemical inertness, thermal stabilization, and resistance to radiation, making them more reliable. The hexagonal (wurtzite) crystal structure of the nitride device leads to unique material properties like large bandgap energies, direct bandgap, high thermal stability, and built-in electric field. The presence of spontaneous and piezoelectric polarization makes them ideal for electronic applications. There has been a great interest in the research of nitride-based quantum well devices in recent years because of the importance of energy conservation and the wide variety of applications, including high-speed and high-power electronics.

The significance of the development and use of III–V nitride-based diodes and FETs for low-loss energy conversion and effective transmission/distribution of electric power capabilities at increased operating temperatures are of great demand. The development of GaN-based electrical switches for fast and efficient operation at a wide range of power levels and compact device sizes, for example, the drastic reduction in gate width and chip area, has created a revolution in the field of power electronics.

2.3.1 Bandgap Engineering

To overcome the material limits of Si and to realize the improvement in device performance needed to meet future necessities, wide bandgap semiconductors such as GaN have come to the forefront due to their superior physical properties.

They offer a higher potential advantage over Si-based devices in power switching and conducting performance, and can withstand high temperatures and high voltages. Figure 2.4 clearly shows the dependence of bandgap and lattice constant of the III–V nitride semiconductor. The bandgap energy (E_g) of GaN (3.4 eV) can be engineered from 2.0 eV to 6.2 eV by adding indium or aluminum ternary material to GaN. Hence the extra degree of freedom is available for researchers to advance bandgap engineering.

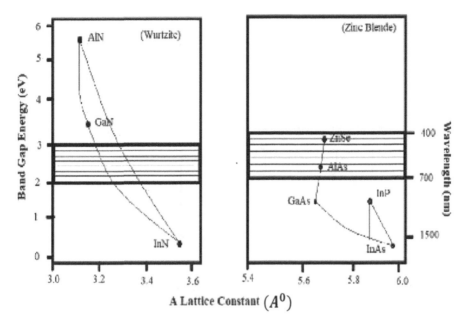

FIGURE 2.4
The comparison and variation of lattice constant vs. bandgap energy vs. wavelength for wurtzite and zinc-blende structure of III–V nitride semiconductors. The bandgap energy (Eg) of GaN
(3.4 eV) is between 2.0 eV and 6.2 eV by adding indium or aluminum ternary material to GaN.

2.3.2 Impact of Polarization in GaN Devices

The strong effects of polarization in III–V nitride heterostructures and their
properties are explored in detail. The unique lattice structure of GaN heterostructures induces polarization effects leading to high carrier concentrations
$n_s \geq 10^{13}$ cm^{-2} without any impurity doping. For AlGaN/GaN HEMTs, spontaneous and piezoelectric polarization influence channel formation and alter
the device output characteristics.

III–V nitride devices have two polarization effects: spontaneous (P_{sp}) and
piezoelectric (P_{pz}) polarization. Spontaneous polarization is due to nitride (N)
crystal and high nitrogen atom electronegativity due to the wurtzite nature.
We know that the wurtzite crystal structure is non-centrosymmetric; it lacks
a center of symmetry, creating a net internal electric field. The effect of piezoelectric polarization is due to the presence of strain and stress effects (i.e., lattice
mismatch) in these heterointerfaces predicted by Bykhovski (see Figure 2.5).

Piezoelectric polarization creates higher carrier concentration and high
electron mobility in the channel (up to $\mu \sim 2000$ cm^2/Vs) without intentional
doping leading to low ON-state resistance, low conduction loss, and higher
converter efficiency. The electrons are properly confined in a quantum well
to avoid alloy impurity scattering. The fast switching properties and direct
bandgap of GaN make it possible to operate at high-frequency and optoelec-
tronic circuits due to its improved electron mobility and dielectric constant.

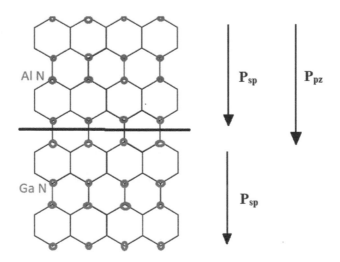

FIGURE 2.5
The output characteristics of AlGaN/GaN HEMTs, and spontaneous and piezoelectric polarization influences the channel formation. III–V nitride devices with two polarization effects, spontaneous (Psp) and piezoelectric (Ppz) polarization.

TABLE 2.2

Comparison of Some Critical Factors on the Fundamental Performance Characteristics of the Devices

Material	Bandgap Energy (eV)	Breakdown Field (MV/cm)	Thermal Conductivity (W/cm*K)	Mobility (cm²/V*s)	Saturated Velocity (*10⁷ cm/s)
Si	1.1	1.5	1.5	1300	1.0
GaAs	1.4	0.5	0.5	6000	1.3
SiC	3.2	4.9	4.9	600	2.0
GaN	3.4	>3.0	2.3	~2000*	2.7

The material properties and their corresponding device and system-level advantages are shown in Table 2.2. Different techniques are developed to design III–V material-based GaN devices with low OFF-state leakage, enhancement mode operation with an improved threshold voltage, drain current density, breakdown for high-power switching, and analog/RF applications.

2.3.3 Lattice Mismatch and Strain in III–V Nitride Semiconductors

Because of the effects of polarization, lattice mismatch, and strain between an epitaxial layer (GaN) and the substrate is essential for analyzing the material quality and crystal stability. The lattice mismatch is calculated from the lattice constants of the epitaxial layers (a_{epi}) and the substrate (a_{sub}) is defined by Equation 2.1

$$\frac{\Delta a}{a_{epi}} = \frac{a_{epi} - a_{sub}}{a_{epi}} \tag{2.1}$$

The commonly used substrates for the fabrication of nitride devices are sapphire (Al$_2$O$_3$), SiC, silicon, and aluminum nitride (AlN). Among these substrates, AlN has a lower lattice mismatch (<1%) in the GaN epitaxial growth, but the high cost and limited wafer size reduced the interest of researchers in the AlN substrate (Hee-Jin Kim et al. 2010). SiC substrate has a reasonably small (3.4%) lattice mismatch compared to the 13% mismatch between the GaN and sapphire substrate. Although a high lattice mismatch is present in silicon, some researchers are using silicon as the substrate material because of its availability and low cost. Even so, the fabrication of silicon-based GaN epitaxial growth requires special attention due to its high lattice mismatching with the GaN epitaxy (Huanyou Wang et al. 2014). Sapphire is the commonly used substrate material for device fabrication because of its comparatively low cost, oxygen erosion resistance, and temperature stability.

2.3.4 AlGaN/GaN Heterostructure HEMTs

Through bandgap engineering, it is possible to continuously vary the bandgap of III–V nitride ternary compound materials by adjusting the In or Al composition in alloys of AlGaN, InAlN, and so on. The typical optimized value of mole fraction 'x' is below 0.18–0.3 (Martin Ko 2003).

$$\text{Al}_x\text{Ga}_{1-x}\text{N 'x' aluminum mole fraction} \begin{cases} x = 0 & \text{AlGaN} \rightarrow \text{GaN} \\ x = 1 & \text{AlGaN} \rightarrow \text{AlN} \\ 0 < x < 1 & \text{AlGaN} \rightarrow \text{Al}_x\text{Ga}_{1-x}\text{N} \end{cases}$$

Al$_x$Ga$_{1-x}$N and Al$_y$In$_{1-y}$N are examples of wide bandgap semiconductors, and GaN and InN are narrow bandgap semiconductors, as shown in Figure 2.6. When a wide bandgap semiconductor is grown on a narrow bandgap semiconductor, the band offset, or conduction band discontinuity is formed

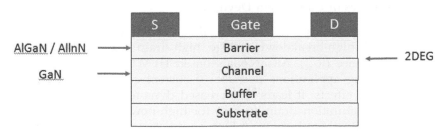

FIGURE 2.6
Cross-sectional view of (In) AlGaN/ GaN heterostructure HEMT and 2DEG formation.

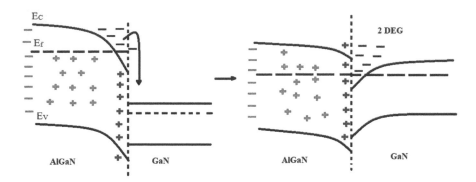

FIGURE 2.7
Energy band of an AlGaN interface with GaN creates different Fermi levels and electron flow
to the GaN channel (left). The whole band energy level of an AlGaN/GaN heterostructure after
Fermi level equalization creates 2DEG in the quantum well (right).

between the interfaces leading to the accumulation of electrons results in
high 2DEG and current in these devices (Asif Khan et al. 1993) in Figure 2.7.
The electron drift from higher bandgap to lower bandgap interfaces is shown
in Figure 2.7.

- Energy bandgap discontinuity highly depend upon band difference
  between valance band ΔE_V and conduction band ΔE_C.
- It also depends on the built-in potential (qV_B).

When the AlGaN/AlInN and GaN layers are grown on top of each other to
create the heterojunction, the atoms at the interface are subjected to mechanical strain (piezoelectric polarization). The electrons are attracted to the
interface by positive polarization charges and form the 2DEG. The electrons
depleted from the wide bandgap layers occur at the narrow bandgap interface, and carriers are confined within the narrow dimension quantum well.
They can move freely within the heterojunction plane and are restricted to a
well-confined space region called the 2DEG, shown in Figure 2.7.

2.3.5 Challenges in GaN-Based Devices

For high-power applications, AlGaN/GaN-based HEMT devices are predominant with high-breakdown voltage, high drain current, and low specific ON-resistance (R_{ON}). AlGaN/GaN-based HEMTs cannot be used for
power switching applications because of current collapse or drain current dispersion effects. It leads to increased dynamic ON resistance and
decreases the saturation drain current for high-power switching applications. Meanwhile, there have been better ways to improve the drain current and reduce current collapse using remarkable growth techniques. The
main reason for the current collapse is the trapping issues near the gate to

the drain region. Primarily, passivation of AlGaN/GaN HEMTs shows better transient response compared with non-passivated devices. Current collapse effects are much degraded with field plate (FP)-structured AlGaN/GaN devices (Figure 2.8). The impact of the FP on current collapse degradation is reviewed by Y. F. Wu et al. (2004).

The FP presence will reduce the dynamic ON-resistance and reduce the trapped charges for higher positive gate bias. We can assume that FP devices are responsible for drain current dispersion reduction and the field-effect recovery process. The trap states in the channel are depleted by AlGaN FP length. The proposed studies state that the current collapse is reduced by a better choice of passivation film and longer FP from the gate to drain in AlGaN devices.

Several efforts have been made to reduce the current collapse dispersion and increased dynamic ON-resistance in AlGaN/GaN HEMT devices, which reduce the trapping issues (Y. Ohno et al. 2004). With the AlN barrier layer in the AlGaN/GaN interface (J. S. Lee et al. 2003), FP techniques (Y. F. Wu et al. 2004), plasma, chemical treatments, and annealing process, the current collapse is minimized (A. P. Edwards et al. 2005). From the literature, the surface states are responsible for current dispersion in GaN-based devices. Also, oxygen impurities are the source of these defective states (P. J. Hansen et al. 1998). In addition to this analysis, several reports show the trap-induced locations and model for trap issues (M. Faqir et al. 2008). Passivation of Si_3N_4 and oxide are the factors to overcome these dispersion and trapping issues (B. Luo et al. 2002). Meanwhile, these issues raise the question about trap sources and 2DEG sources. Passivation effects show that increasing the drain current will increase 2DEG carriers and the surface charge density. So the interface charge is considered neutral. It leads to increased positive charges at the dielectric/AlGaN region (S. Arulkumaran et al. 2004).

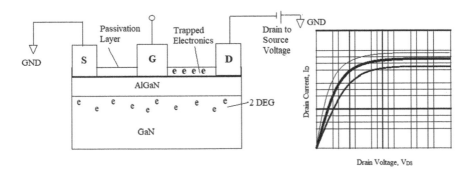

FIGURE 2.8
The degraded effects with field plate (FP)-structured AlGaN/GaN devices and transient response comparison of non-passivated devices with AlGaN/GaN HEMTs. The characteristics of drain voltage vs. drain current; when drain voltages change to maximum, then the current will be saturated.

A few explanations for the passivation effects in the device are reported. Initially, the passivation process increases the density of states in the device. Another reason says that the passivation layer makes trapped electrons inaccessible from the gate side (R. Vetury et al. 2001). Meanwhile, a third explanation says that the passivation layer reduces the peak electric field near the gate regions.

The peak electric field near the gate-to-drain region is assumed as a dominant factor for trap occupation near the gate region. The FP in HEMTs increases the breakdown voltage, and at the same time, it reduces the effect of the peak electric field near the gate edge (O. Mitrofanov and Manfra 2003). Very little experimental evidence proves the suppression of the density of states and trap charges near the gate region. The simulation model for field-plated device characterization is essential to overcome these distortions.

2.3.6 AlGaN Channel HEMT Device for High-Power Applications

RF power enhancement is the primary goal of AlGaN/GaN channel HEMT devices. The enhancement of breakdown voltage is essential to increase the power density. Employing AlN content in GaN leads to increased breakdown voltage because of the higher electric field and higher bandgap in AlN-based devices. The expected breakdown field of AlN is four times greater than in GaN-based devices (A. Raman et al. 2008). So Al incorporation in GaN-based devices has become a promising alternative for high-power devices. Thermal conductivity is an additional advantage in AlN-based devices. Mobility degradation is a limiting factor in AlN devices because of alloy scattering.

Very recently, adding Al to the GaN channel became very attractive because of higher breakdown voltage enhancement and high thermally stable device fabrication (A. Raman et al. 2008). The most challenging part of high-power device fabrication is achieving a low resistive ohmic contact. Simultaneously, with AlGaN channel devices, higher breakdown voltage enhancement is achieved four times compared with GaN-based channel devices. The drain voltage of more than 30 V is difficult in AlGaN channel devices. Nanjo et al. demonstrated that high-breakdown voltage and high carrier density achieved with the AlGaN channel are limited because of high ohmic resistance. Si ion implantation doping can solve these issues. By employing this fabrication process, the device can achieve low ohmic contact resistance, high drain current density, and extremely high OFF-state breakdown voltage (T. Nanjo et al. 2007). Till now, only a few possible OFF-state breakdown voltage enhancements were reported for AlGaN channel HEMT devices. The direct current (DC) characteristics of AlGaN channel HEMT are employed using field plate electrodes.

2.3.7 Effect of FP on AlGaN Channel HEMT Devices

The OFF-state breakdown voltage characterization is well defined in AlGaN-based devices. Adding FP in these devices is an additional advantage to achieve higher breakdown voltage (Eldad Bahat-Treidel et al. 2010). This FP distributes the peak electric field from the gate-to-drain region. This distribution can achieve a higher breakdown compared with conventional devices.

The reduction of gate leakage current achieves better performance. The multi-FP device generates a single peak under the gate edge of the AlGaN device; thus, it distributes the electric field equally throughout the device. These devices exhibit higher efficient breakdown voltage compared with a single-FP device. The main disadvantages of this device are the limitation of device capacitance. These effects limit the frequency and switching performance of the device. More optimization or research is needed to achieve better switching performance (W. Saito et al. 2005). The additional advantages of multi-gate field plates (MGFPs) are the better gate-to-drain control compared with conventional device structures. The characterization of AlGaN channel devices is performed by technology computer-aided design (TCAD) simulation.

Figure 2.9 shows that an AlGaN channel and back-barrier (BB) were considered along with MGFPs. It was predicted that the AlGaN channel device and the presence of BBs would increase OFF-state breakdown (Bahat-Treidel et al. 2007). Here, we use the defect-free polarization-induced depletion AlGaN BBs layers, and the effects are visualized.

Figure 2.10 explores the effect of percentage composition of aluminum back barrier on break down voltage for different materials (Si, SiC, GaN).

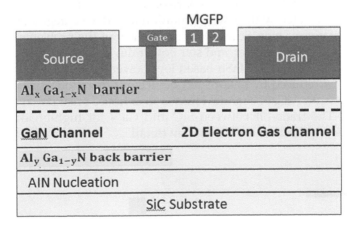

FIGURE 2.9
Cross-sectional view of multiple-gate field-plated AlGaN/GaN channel device with field plate addition to achieve higher breakdown voltage compared with conventional devices.

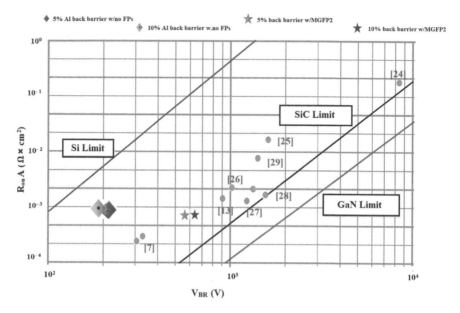

FIGURE 2.10
Effect on Al back-barrier in Si, SiC, and GaN of 5% and 10% with and without FP on breakdown voltage.

2.4 Summary

This chapter provides a brief comparison of III–V heterostructures to sort the issues associated with power limitation and device degradation issues. Moreover, it gives a significant literature review of the problems associated with low-power, high-power, and self-power devices. Threading dislocation analysis is the solution for InSb-based low-power devices. GaN degradation issues can be reduced by field-plated device optimization. Also, the effects of field plates on AlGaN channel HEMTs to enhance breakdown voltage are discussed. The trade-off between SiC and GaN for high-breakdown and high-power applications is discussed in detail.

References

S. Arulkumaran, T. Egawa, H. Ishikawa, T. Jimbo, and Y. Sano, "Surface passivation effects on AlGaN/GaN high-electron-mobility transistors with SiO2, Si3N4, and silicon oxynitride," *Appl. Phys. Lett.*, vol. 84, pp. 613–615, 2004.

M. Asif Khan, A. Bhattarai, J. N. Kuznia, and D.T. Olson, "High electron mobility transistor based on a GaN/Al$_x$Ga$_{1-x}$N hetero-junction," *Appl. Phys. Lett.*, vol. 63, no. 9, pp 1214–1215, 1993.

Eldad Bahat-Treidel, Oliver Hilt, Frank Brunner, Victor Sidorov, Joachim Würfl, and Günt her Tränkle, "AlGaN/GaN/AlGaN DH-HEMTs breakdown voltage enhancement using multiple grating field plates (MGFPs)," *IEEE Trans. Electron Devices*, vol. 57, no. 6, June 2010.

E. Bahat-Treidel, V. Sidorov, J. Würfl, and G. Trankle, "Simulation ofAlGaN/GaN HEMTs' breakdown voltage enhancement using grating field plates," in Grasser T., Selberherr S. (eds), *Simulation of Semiconductor Processes and Devices*, 2007. Springer, Vienna. https://doi.org/10.1007/978-3-211-72861-1_66, pp. 277–280.

S. Dasgupta Raman, S. Rajan, J. S. Speck, and U. K. Mishra, "AlGaN channel high electron mobility transistors: device performance and power-switching figure of merit," *Jpn. J. Appl. Phys.*, vol. 47, no. 5, pp. 3359–3361, May 2008.

P. Edwards, J. A. Mittereder, S. C. Binari, D. S. Katzer, D. F. Storm, and J. A. Roussos, "Improved reliability of AlGaN-GaN HEMTs using an NH3/plasma treatment prior to SiN passivation," *IEEE Electron Device Lett.*, vol. 26, pp. 225–227, 2005.

M. Faqir, G. Verzellesi, A. Chini, F. Fantini, F. Danesin, G. Meneghesso, E. Zanoni, and C. Dua, "Mechanisms of RF current collapse in AlGaN/GaN High Electron Mobility Transistors," *IEEE Trans. Device Mater Reliab.*, vol. 8, pp. 240–247, 2008.

P. J. Hansen, Y. E. Strausser, A. N. Erickson, E. J. Tarsa, P. Kozodoy, E. G. Brazel, J. P. Ibbetson, U. Mishra, V. Narayanamurti, S. P. DenBaars, and J. S. Speck, "Scanning capacitance microscopy imaging of threading dislocations in GaN films grown on (0001) sapphire by metal-organic chemical vapor deposition," *Appl. Phys. Lett.*, vol. 72, pp. 2247–2249, 1998.

J. W. Johnson Luo, J. Kim, R. M. Mehandru, F. Ren, B. P. Gila, A. H. Onstine, C. R. Abernathy, S. J. Pearton, A. G. Baca, R. D. Briggs, R. J. Shul, C. Monier, and J. Han, "Influence of MgO and Sc2O3passivation on AlGaN/GaN high-electron-mobility transistors," *Appl. Phys. Lett.*, vol. 80, pp. 1661–1663, 2002.

C. J. Kiely, J.-I. Chyi, A. Rockett, and H. Morkoc, "On the microstructure and interfacial structure on InSb layers grown on GaAS(100) by molecular beam epitaxy," *Philos. Mag. A*, vol. 60, pp. 321, 1989.

Hee-Jin Kim, Suk Choi, Seong-Soo Kim, Jae-Hyun Ryou, Douglas Yoder, P, Russell D. Dupuis, Alec M. Fischer, Kewei Sun, and Fernando A. Ponce, "Improvement of quantum efficiency by employing active-layer-friendly lattice-matched InAlN electron blocking layer in green light-emitting diodes," *Appl. Phys. Lett.*, vol. 96, no. 10, p. 101102, 2010.

P. B. Klein, and S. C. Binari, "Photoionization spectroscopy of deep defects responsible for the current collapse in nitride-based field-effect transistors," *J. Phys. Cond. Matt.*, vol. 15, p. R1641, 2003.

Martin Ko, "AlGaN/GaN MBE 2DEG heterostructures: interplay between surface, interface and device-properties," pp. IV, 213s, 16, 2003.

J. S. Lee, J. W. Kim, J. H. Lee, C. S. Kim, J. E. Oh, and M. W. Shin, "Reduction of current collapse in AlGaN/GaN HFETs using AlN interfacial layer," *Electron. Lett.*, vol. 39, pp. 750–752, 2003.

Denis Mamaluy, and Xujiao Gao, "The fundamental downscaling limit of field-effect transistors," *Appl. Phys. Lett.*, vol. 106, p. 193503, 2015. DOI:http://dx.doi.org/10.1063/1.4919871.

O. Mitrofanov, and M. Manfra, "Mechanisms of gate lag in GaN/AlGaN/GaN high electron mobility transistors," *Superlattices Microstruct.*, vol. 34, pp. 33–53, 2003.

T. Nanjo, M. Takeuchi, M. Suita, Y. Abe, T. Oishi, Y. Tokuda, and Y. Aoyagi, "Remarkable breakdown voltage enhancement in AlGaN channel HEMTs," *Proc. IEEE IEDM*, 2007, pp. 397–400.

Y. Ohno, T. Nakao, S. Kishimoto, K. Maezawa, and T. Mizutani, "Effects of surface passivation on the breakdown of AlGaN/GaN high-electron-mobility transistors," *Appl. Phys. Lett.*, vol. 84, pp. 2184–2186, 2004.

W. Saito, M. Kuraguchi, Y. Takada, K. Tsuda, I. Omura, and T. Ogura, "Design optimization of high breakdown voltage AlGaNGaN power HEMT on an insulating substrate for R ONA – VB trade-off characteristics," *IEEE Trans. Electron Devices*, vol. 52, no. 1, pp. 106–111, Jan. 2005.

R. Vetury, N. Q. Zhang, S. Keller, and U. K. Mishra, "The impact of surface states on the DC and RF characteristics of AlGaN/GaN HFETs," *IEEE Trans. Electron Devices*, vol. 48, pp. 560–566, 2001.

Huanyou Wang, Yalan Li, and Penghua Zhang, "Enhancement of InGaN-based light emitting diodes performance grown on cone-shaped pattern sapphire substrates," *J. Mater. Sci. Chem. Eng.*, vol. 2, no. 7, pp. 53–58, 2014.

Y. F. Wu, A. Saxler, M. Moore, R. P. Smith, S. Sheppard, P. M. Chavarkar, T. Wisleder, U. K. Mishra, and P. Parikh, "30-W/mm GaN HEMTs by field plate optimization," *IEEE Electron Device Lett.*, vol. 25, pp. 117–119, Mar. 2004.

M. J. Yang, B. R. Bennett, M. Fatemi, P. J. Lin-Chung, W. J. Moore, and C. H. Yang, *J. Appl. Phys.*, vol. 87, p. 8192, 2000.

M. Yano, T. Takase, and M. Kimata, "Heteroepitaxial InSb films grown by molecular beam epitaxy," *Phys. Status Solidi A*, vol. 54, no. 707, 1979.

X. Zhang, A. E. Staton-Bevan, and D. W. Pashley, "A TEM investigation of the initial stages of InSb growth on GaAS (001) by molecular beam epitaxy," *Mater. Sci. Eng. B*, vol. 7, no. 3, pp. 203–208, 1990.

3

III–V Heterostructure Devices for High-Frequency Applications

R. Saravanakumar

CONTENTS

3.1 Introduction

Presently, GaAs-based heterostructures are well recognized for microwave circuit applications (J. W. Lee et al. 1998). The pseudomorphic high electron mobility transistor (pHEMT) combination of GaAs structures like AlGaAs/InGaAs/GaAs (Y. Bito et al. 2001), GaAs/InGaAs/GaAs (P. H. Lai et al. 2006), and InGaP/InGaAs/GaAs (Y. S. Lin et al. 2005) offer high power, high frequency, and high mobility compared to AlGaAs/GaAs heterostructures. The high bandgap discontinuity (E_c) of InGaAs channel material with an AlGaAs barrier enhances the two-dimensional electron gas (2DEG) carrier density in the channel because of the low effective mass (Y. S. Lin et al. 2006). However, the indium content in GaAs creates dislocation in the channel due to the high lattice mismatch with the barrier. The high indium mole fraction induces strain relaxation between AlGaAs/InGaAs layers by creating a higher dislocation. Incorporating the interfacial layer between the barrier and channel layer reduces strain in the device without creating misfit dislocation and enables easy growth of high-quality heterostructures (J. W. Matthews 1975).

DOI: 10.1201/9781003093428-3

By introducing an additional δ-doping concentration in the spacer layer above the barrier layer, the device carrier density and mobility further increase because of the high 2DEG concentration induced in the channel. At the same time, the breakdown voltage of the device will decrease drastically. There should be a compromise between breakdown voltage and current density formation at the channel for optimal device performance. Various structures are proposed to enhance the breakdown voltage without degrading the current density (Y. Okamoto et al. 1995). A high carrier density is obtained by using δ-doping; however, the silicon auto compensation is much reduced by the application of active Si concentration in the δ-doped region (E. F. Schubert et al. 1987). Overhomogeneously doped HEMTs and δ-doped HEMTs have several advantages like less parallel conduction, enhanced mobility, high 2DEG concentration, high-breakdown voltage, and reduced leakage current (W. S. Lour et al. 2001).

3.2 Device Description

The device characteristics of pHEMTs with δ-doped and undoped conditions are analyzed. The power-added efficiency, output power, and power gain are analyzed using TCAD Sentaurus tools. The device structure shown in Figure 3.1 is comprised of a 200 nm buffer layer followed by three layers of InGaAs material with different compositions. The bottom 10 nm thick $In_xGa_{1-x}As$ layer with x = 0.53 represents the channel, the middle 5 nm thick $In_xGa_{1-x}As$ layer with x = 0.22 represents the reduced impurity scattering from the barrier, and the top 50 nm thick $In_xGa_{1-x}As$ layer with x = 0.22 represents the barrier. A 500 nm thick $Al_xGa_{1-x}As$ with x = 0.3 is used as a cap region to reduce the lattice mismatch and improve gate control. The cap region also reduces the source and drain ohmic contact resistance. The ohmic region is n-doped with a concentration of $1 \times 10^{19}/cm^3$. The AlGaAs/InGaAs/InGaAs/GaAs pHEMT is also analyzed for two different δ doping concentrations: low and high. The effect of power-added efficiency and gain is also analyzed with and without δ-doped regions.

3.3 Power Gain, Output Power, and Power-Added Efficiency

The microwave power measurement of power gain (PG), output power (OP), and power-added efficiency (PAE) is analyzed using technology computer-aided design (TCAD) simulation tools. The measurement is done with input power varying from –20 to 10 dBm at a frequency range of 2.4 GHz. From

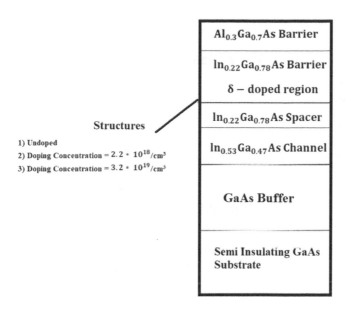

Structures

1) Undoped
2) Doping Concentration = $2.2 * 10^{18}/cm^3$
3) Doping Concentration = $3.2 * 10^{19}/cm^3$

$Al_{0.3}Ga_{0.7}As$ Barrier
$In_{0.22}Ga_{0.78}As$ Barrier
δ – doped region
$In_{0.22}Ga_{0.78}As$ Spacer
$In_{0.53}Ga_{0.47}As$ Channel
GaAs Buffer
Semi Insulating GaAs Substrate

FIGURE 3.1

The influence of device structure with different doping concentrations. As the contact voltage decreases, the doping concentration will increase due to carrier concentration. For low power output, the trade-off between the doping concentration on current and voltages will happen.

Figures 3.2 and 3.3, the 23 dB power gain ($G = P_{out}/P_{in}$) is achieved at a small-signal regime. Also, the saturated output power measured 19 dB with a matched gain of 10 dB.

The frequently used drive power at microwave frequencies is measured using PAE. It is used for analyzing power-added performance when the gain is high. It is an essential parameter for radio frequency (RF) power amplifiers and low-power devices as the cooling system cost is significant compared to actual equipment. The equation used for measuring PAE and overall efficiency is

$$PAE = (Pout - Pin) / Pdc = \eta(1 - 1/G) \tag{3.1}$$

$$P_{overall} = Pout / (Pdc + Pin) \tag{3.2}$$

3.4 Review of InGaAs and InAs HEMTs for High-Frequency Applications

III–V material devices such as InGaAs- and InAS-based MOSFET/HEMT devices with a single gate (SG)/double gate (DG) pave the way for the

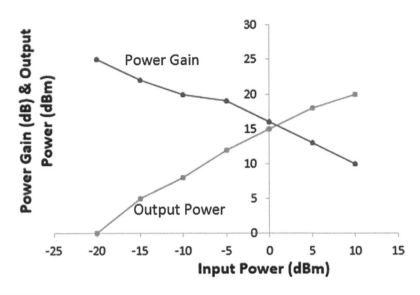

FIGURE 3.2
pHEMT with load pull data simulation measurement at 500 nm gate length and 100 μm width. A 2.4 GHz frequency is used for measurement with the off-state constant bias of V_{gs} at 0 V and V_{ds} at 4 V.

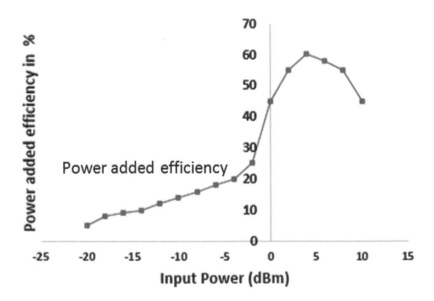

FIGURE 3.3
Load pull data simulation of PAE with input power sweep from −20 dBm to 10 dBm.

betterment of DC and RF performance along with low short-channel effects. The influence of new techniques like gate channel and barrier engineering significantly improve the performance of the device.

Wu et al. (2009) fabricated a 150 nm gate length inversion-mode InGaAs channel MOSFET with Al_2O_3 as the gate dielectric material. The channel length varied from 150 to 250 nm, and device metrics such as transconductance, subthreshold slope, drain-induced barrier lowering, and threshold voltages are analyzed. The transconductance was improved from 1.1 to 2.5 mS/μm to reduce Al_2O_3 gate dielectric thickness from 5 to 2.5 nm. The reduction of the gate dielectric thickness improves the analog performances of the device.

Fei Xue et al. (2011) described an InGaAs device with a buried channel MOSFET, with enhanced performance by inserting the InAs layer between the channel material. A comparison of device parameters, such as pure InGaAs and InAs inserted channel material, is made. The results imply that the InGaAs/InAs/InGaAs channel layer has an excellent off-state and saturation property and improved mobility up to 37% compared to the pure InGaAs channel material. Moreover, InGaAs MOSFET device performance is observed by varying the indium compositions of channel material such as 53% and 70%, respectively. It is found that higher transconductance and lower subthreshold swing are achieved for x = 0.53 In mole fraction.

Alireza Alian et al. (2013) explained the concept of different channel thickness with respect to InGaAs channel MOSFETs. The channel thickness varied from 3 to 10 nm and found an optimum channel thickness of 10 nm. As the channel thickness is reduced, the mobility degrades and the accumulation capacitance value increases due to the variation in the inverse charge profile.

Suchismita Tewari et al. (2013) introduced different barrier layers such as InP and InAlAs in InGaAs channel MOSFET devices to improve the electron mobility for high-performance logic applications. The device parameters such as transconductance, transconductance-to-drive current ratio, drain conductance, intrinsic gain, and cutoff frequency (f_T) were compared for single (InP) and double (InP/InAlAs) barrier layers. The results imply that the double barrier layer exhibits a better analog performance compared to the single barrier. The impact of indium content in the channel significantly improved the RF performance metrics with increased short-channel effects.

Dae-Hyun Kim and Jesús A. del Alamo (2008) analyzed the impacts of lateral and vertical scaling of $In_{0.7}Ga_{0.3}As$ HEMTs. They synthesized the behavior of InGaAs HEMTs in terms of gate length and insulator thickness. The authors suggested that better short-channel behavior and electrostatic integrity can be achieved with a gate length of 60 nm and by reducing the insulator thickness of $In_{0.52}Al_{0.48}As$. Simulation results were also discussed, implying various factors such as gate delay of CV/I = 1.6 ps, SS = 88 mV/V, V_T = −0.02 V, I_{ON}/I_{OFF} = 1.6 × 10^4, and DIBL = 93 mV/V at V_{DD} = 0.5 V. In addition to this, the superior performance of f_T = 562 GHz was achieved with an insulator thickness of 4 nm on a 25 nm gate-length device. It was concluded that insulator thickness scaling plays a vital role in attaining sub-100 nm gate-length InGaAs HEMTs, which also has a trade-off in parasitic source resistance (R_s) and gate leakage current (I_G).

Akira Endohet al. (2009) reported that multicap structures significantly reduce the source and drain parasitic resistance and fabricated 30 nm gate pseudomorphic $In_{0.52}Al_{0.48}As/In_{0.7}Ga_{0.3}As$ HEMTs on InP substrate. The various DC and RF characteristics are observed at different temperatures, such as 1677 and 300 K, with multiple bias conditions. The maximum cutoff frequency of 600 GHz was achieved and reduced to 498 GHz at 300 K. InAs and InSb channel layers with high electron mobility and high velocity to conduct low-voltage low-power operations were compared to $In_xGa_{1-x}As$. The cryogenic InP-based ultrashort-gate pseudomorphic $In_{0.52}Al_{0.48}As/In_{0.7}Ga_{0.3}As$ HEMT is an essential device for low-power, low-voltage, and high-speed applications.

To overcome the drawbacks related to the device scaling, Tae-Woo Kim et al. (2009) proposed a 30 nm $In_{0.7}Ga_{0.3}As$ inverted HEMT structure, which includes the barrier layers at the top and bottom, leading to low leakage current along with better scalability, logic performance, and high-frequency characteristics (at f_T = 500 GHz and f_{max} = 550 GHz) at L_g = 30 nm. They also suggested that an inverted HEMT is similar to III–V compound material designs, but scaling to very low dimensions is possible when a high K-dielectric acts as a gate insulator.

Dae-Hyun Kim et al. (2010) describe 50 nm enhancement-mode (E-mode) $In_{0.7}Ga_{0.3}As$ pHEMTs with maximum frequency 1 THz. They also stated that it is challenging to achieve the E-Mode pHEMT due to the tight barrier control and gate length approaching the sub-50 nm regime. They explained the platinum gate sinking process along with a thin barrier layer of $In_{0.52}Al_{0.48}As$. Fabricated devices with L_g = 50 nm imply the transconductance, threshold voltage, f_T, and f_{max} of 1.75 mS/µm, 0.1 V, 465 GHz, and 1.06 THz, respectively, at a moderate value of V_{ds} = 0.75V. These devices have a sharp subthreshold characteristic assessed by a SS = 80 mV/dec, and the drain-induced barrier lowering value is 80 mV/V.

Akira Endoh et al. (2010) demonstrated the three-valley model (Γ, L, X), which was carried out through Monte Carlo (MC) simulation for the band structures of $In_{0.75}Ga_{0.25}As$ and $In_{0.52}Al_{0.48}As$ where the energy bandgap is higher in the $In_{0.75}Ga_{0.25}As$ channel due to compressive strain. The various scattering mechanisms such as acoustic phonon scattering, ionized impurity scattering, alloy scattering, and so on, and ionization impact were considered in the simulations. A two-dimensional Monte Carlo simulation was carried out on 200 nm gate InP-based $In_{0.52}Al_{0.48}As$ /$In_{0.75}Ga_{0.25}As$ HEMTs at 300 and 16 K to elucidate the consequence of temperature on electron transport. As the temperature is decreased, there is an increase in transconductance and cutoff frequency from 1110 mS/mm at 300 K to 1400 mS/mm at 16 K and from

168 GHz at 300 K to 223 GHz at 16 K, respectively. This has led to an increase in electron velocity from 3.60×10^7 cm/s at 300 K to 5.26×10^7 cm/s at 16 K.

Tae-Woo Kim et al. (2010) introduced a new self-aligned gate technology along with non-alloyed Mo-based ohmic contacts for InGaAs HEMTs, a vital role for achieving a low parasitic capacitance. This can be achieved by shrinking the source-gate contact separation (L_{GS}), which paves the way for improving III–V device frequency response. By this new process, a gate-to-source contact separation (L_{GS}) of 100 nm and side recess length (L_{side}) of 200 nm are chosen with a contact resistance of 7 ohm-μm and a source resistance of 147 ohm-μm. Molybdenum-based ohmic contacts show an excellent thermal stability up to 600°C for 60 nm gate length self-aligned InGaAs HEMTs with $g_m = 2.1$ mS/μm at $V_{DS} = 0.5$ V, and $f_T = 580$ GHz and $f_{max} = 675$ GHz at $V_{DS} = 0.6$ V.

D. Saguatti et al. (2010) demonstrated InAlAs-InGaAs HEMTs, which fetch the best frequency and low noise performance, and the drawback of stating the low breakdown voltage, resulting from a small critical electric field on the channel. Hence, to achieve a high-power density and high-efficiency power amplifiers (PAs) design, a high-voltage operation is required, achieved by the field plate (FP) structures. It consists of an extension of the gate electrode toward the drain contact. This structure effectively reduces the electric field magnitude at the drain side of the gate edge, leading to the improvement of maximum operating voltage, which must be traded off with the decrease in RF performance associated with the additional gate–drain capacitance. A high-voltage InGaAs-InAlAs HEMT using a 2D hydrodynamic device and TCAD approaches are adopted for this design. Comparison results in terms of DC and RF performance are made by applying the InP HEMT technology incorporating field plate structures with optimal length and passivation thickness. HEMT design with the field plate technique makes feasible the design of high-voltage, high-efficiency power amplifiers, and robust low-noise amplifiers.

M. Ahmad et al. (2012) focused on incorporating a highly strained depletion-mode $In_{0.7}Ga_{0.3}As$ channel pseudomorphic HEMT and simulated the same using SILVACO. The effects of barrier thickness, the impact of delta doping, and gate length (L_g) scaling were provided in this work. The variation of barrier layer thickness affects the electric field in the device and alters overall device performance. Depending on the position and amount of delta doping level, the threshold voltage varies from a negative level to a positive. The transconductance initially increases with an increase in delta doping but then tends to saturate after the doping reaches a value of 3×10^{19} cm^{-2}. The variation of the gate length affects the drain current and the RF performance of the device. The authors were specific in pointing out the dimensions such as gate length of the pHEMT; source-to-gate spacing; and delta doping as 1.2 μm, 2 μm, and 2.8×10^{19} cm^{-2}, respectively. The simulation results show optimized performance in terms of transconductance, pinch-off voltage, and maximum drain current.

Zhong Yinghui et al. (2012) proposed a model of 88 nm gate length InP-based InAlAs/InGaAs HEMTs with the maximum oscillation frequency and cutoff frequency as 185 GHz and 100 GHz, respectively. The RF characteristics of HEMTs with respect to various L_{side} of 1070, 412, and 300 nm at constant bias conditions were explored. The kink effect plays a vital role in DC characteristics when the dimensions are increased on the L_{side}. Similarly, f_T and f_{max} increase reaching the point of 100 GHz and 215 GHz. L_{side} = 412 nm and L_{side} dependence are mainly briefed in parasitic resistance and parasitic capacitance R_d and R_s. In addition to this, as C_{gd} is larger, the shorter will be the L_{side}, and at the same time, f_{max} will be reduced with an increase in C_{gd}. Further, the transconductance and f_T can be increased with a decrease in the recess region, decreasing the access part of R_s and R_d.

Wang Li-Dan et al. (2014) described the fabrication technique, design procedure, and the characteristics for 100 nm gate length $In_{0.52}Al_{0.48}As/In_{0.53}Ga_{0.47}As$ InP-based HEMTs along with an asymmetrically recessed gate with drain-side lateral recess length L_{rd} = 404 nm and source-side lateral recess length L_{rs} = 550 nm leading to the balance of f_T and f_{max} along with maximum transconductance $g_{m.max}$. The off-state breakdown voltage has been improved to 5.92 V along with a high f_T/f_{max} of 249/415 GHz and maximum extrinsic transconductance by increasing the distance between the gate and drain. The challenging applications with respect to this performance are highlighted in power amplifiers, low-noise amplifiers, and high-speed circuits.

Lisen Zhang et al. (2014) explained the fabrication of a T-shaped gate with a stem height of 200 nm and source-drain spacing 2 μm, gate width of 2 × 20 μm, and 70 nm gate length $In_{0.7}Ga_{0.3}As/In_{0.52}Al_{0.48}As$ on InP-based HEMT, which minimizes the parasitic capacitances. The simulation results in maximum peak extrinsic transconductance, a maximum drain current density, maximum oscillation frequency, and cutoff frequency are discussed.

Mohanbabu et al. (2017) investigated the impact of channel thickness (T_{CH}) and the successful enhancement of device performance under normal off operation by introducing the $HfO_2/GaN/InN/GaN/In_{0.9}Al_{0.1}N$ heterostructure metal insulator–semiconductor HEMT (MISHEMT). A significant increase in n_{2DEG}, I_{ON}, g_m, f_T, and f_{max} demonstrates the potential of E-mode N-polar InN channel MISHEMTs. Also, the inclusion of the lattice-matched $In_{0.9}Al_{0.1}N$ back-barrier in N-polar device technology could deliver more adaptability and advancement of E-mode designs. As T_{CH} becomes thinner, the device's analog/RF performance is enhanced, and impressive peak f_T/f_{max} values of 90 GHz/109 GHz and 180 GHz/260 GHz are fetched, respectively, under the condition of L_g = 100 nm at V_{ds} = 0.5 V and 1 V. It can be a favorable device design for future analog/RF and power amplifier applications.

3.5 Review of InAs-Based Composite Channel HEMT Devices

Dae-Hyun Kim and Jesús A. del Alamo (2010) demonstrated 30 nm gate-length devices with a thin $In_{0.52}Al_{0.48}As$ barrier and InAs channel layer, showing excellent DC and RF characteristics such as 80 mV of threshold voltage, 1.83 mS/μm of transconductance, 85 mV/V of DIBL, 601/609 GHz of cutoff/maximum oscillation frequency, and 2.5×10^7 cm/s of injection velocity at $V_{DS} = 0.5$ V. In this design, short-channel effects are very much reduced due to buried Pt gate and thin barrier thickness.

Tae-Woo Kim and Jesús A. del Alamo (2011) extracted the injection velocity of InAs HEMTs with a 5 nm thick channel and compared them with the 10 nm thicker channel. The results show that as the channel thickness is scaled down, the electron mobility is degraded due to the small injection velocity reduction. A 40 nm gate length and 5 nm thin InAs channel device measured an injection velocity of 3.3×10^7 cm/s. The results suggest that ultrathin InAs channel quantum-well FETs have the capability of downsizing to very short dimensions.

Gomes et al. (2012) demonstrated the performance of narrow-band semiconductor channel material such as InAs, InSb, and $In_{0.53}Ga_{0.47}As$ for future high-speed and low-power applications. It is observed that the $In_{0.53}Ga_{0.47}As$ channel material suffered from lower short-channel effects than the InAs and InSb channel materials. Moreover, InSb material showed a high drain current, high RF performance, and lower intrinsic delay with a lower I_{ON}/I_{OFF} ratio with high leakage current.

Faiz Fatah et al. (2012) described large indium content in the channel, which improved the device performance in all aspects. Yet, finding the optimum bias condition of this narrow-band material is difficult. Indium material is affected by impact ionization at high drain bias conditions degrading the overall device performance in terms of severe kink effect, peak output conductance, current instability, increased gate current, low breakdown voltage, and more channel noise. In this work, the effect of impact ionization of 40 nm gate-length InAs channel HEMT under different bias conditions fetched a cutoff frequency of 615 GHz when the biasing voltage is before the appearance of impact ionization. Simultaneously, suppose the drain bias voltage is increased above the specific level. In that case, the total gate capacitance value is increased due to impact ionization and degrades the device cutoff frequency.

Edward-Yi Chang et al. (2013) explained various factors that improve the RF characteristics of HEMT devices, and they are (a) scaling the gate length, (b) reducing the source-to-drain spacing, (c) increasing the indium content in the channel material, and (d) reducing the barrier/channel thickness. These factors minimize parasitic resistance/capacitance, decrease short-channel effects, and improve the channel electrostatic integrity. Further, the RF performance can be enhanced by the buried platinum gate and high stem height. The buried platinum gate reduces the gate-to-channel distance; hence, it improves the channel aspect ratio, and the high stem height reduces the parasitic resistance/capacitance of the gate. The device was fabricated at L_g = 60 nm, T_B = 5nm, T_{CH} = 5nm, and stem height = 250 nm, with transconductance of 2114 mS/mm, cutoff frequency of 710 GHz, and maximum frequency of oscillation 478 GHz at V_{ds} = 0.5 V. The cutoff frequency is much deviated from the maximum frequency of oscillation due to the narrow L_{side}. This result shows that this device is suitable for submillimeter wave applications.

3.6 Review of InGaAs/InAs Channel Double-Gate MOSFET/HEMT Devices

N. Mohankumar et al. (2010) demonstrated the impact of channel and gate engineering on the analog and RF performances of DG-MOSFET with dual-metal gate technology. The dual-metal DG-MOSFET showed better performance compared to the single-metal DG-MOSFET in terms of the improved gain of 45% by gate engineering and 35% by channel engineering. Through the gate engineering technique, f_T and f_{max} values have increased to 21.6% and 20%, respectively. With the channel engineering technique, f_T was reduced by 2.7%, and the f_{max} is nearly equal. The authors also suggested that the dual-metal gate technology is a suitable technique for low-power subthreshold analog performances and high-power RF applications.

Ge Ji et al. (2012) developed a physical model based on hydrodynamic simulation for scaling and improving InGaAs/InP double heterojunction bipolar transistors (DHBTs). The model depends on the hydrodynamic equation, which can precisely describe nonequilibrium conditions like quasiballistic transport in the thin base and the velocity overshoot effect in the depleted collector. Furthermore, the model includes few physical effects such as bandgap narrowing, effective variable mass, and doping-dependent mobility at high fields. The measured and simulated values of f_T and f_{max} are well matched for lateral and vertical device scaling. The model derived in this work is well suited for designing InGaAs/InP DHBTs.

Neha Verma et al. (2013) investigated quantum mechanics effects in modeling DG InAlAs/InGaAs/InP HEMT at L_g = 50 and 100 nm. The density gradient model analyzes DG InGaAs channel HEMT device carrier concentration, and the device characteristics such as drain current, transconductance, output conductance, gate capacitance, and cutoff frequency of DG InAlAs/InGaAs/In pHEMT devices are computed. These results are compared with the semiclassical model and experimental results, and the comparisons showed good agreement with each other.

Hojjatollah Sarvari et al. (2016) presented a report on short-channel effects of SG and DG graphene nanoribbon field-effect transistors (GNRFETs). The author analyzed the short-channel parameters such as V_T, I_{ON}, I_{OFF}, SS, and DIBL versus channel length and oxide thickness of GNRFETs. The gate capacitance, transconductance, cutoff frequency, and switching delay are calculated for single- and double-gate devices. Furthermore, the effects of doping in the channel on the threshold voltage and the frequency response of double-gate GNRFETs are discussed. The results give an excellent insight into the devices and are very useful for digital applications.

3.7 Summary

In this chapter, the DC characteristics of δ-doped AlGaAs/InGaAs/ InGaAs/ GaAs pHEMTs and high power measurements for a gate width of 100 μm were reviewed. In the DC characteristics analysis for the high δ-doped structures, a maximum drain current of 550 mA/mm, transconductance of 220 mS/mm, gate-to-drain breakdown voltage of 32.2 V, and carrier density of $4.2 \times 10^{12}/cm^2$ were achieved. This device provided a minimum noise figure of 0.79 dB with a gain of 17 dB at 2.4 GHz. The device also showed a 23 dB power gain, an output power of 19 dB, and maximum power-added efficiency of 63%. This literature review has paved the way for identifying the problem and the essential parameters required to meet the goal. Some of the identified factors necessary to achieve better DC and RF performance are optimum gate length, a recessed gate structure, and the concept of reducing the barrier and channel thickness. In addition to this, analyzing and selecting the appropriate III–V compound material such as InGaAs/ InAs, which is most suitable for high-frequency applications, are discussed.

References

M. Ahmad, H..T. Butt, T. Tauqeer, and M. Missous, 2012, "DC Characterization of InGaAs/InAlAs/InP Based Pseudomorphic HEMT," The Ninth International Conference on Advanced Semiconductor Devices and Microsystems, pp. 187–190.

Alireza Alian, Mohammad Ali Pourghaderi, Yves Mols, Mirco Cantoro, Tsvetan Ivanov, Nadine Collaert, and Aaron Thean, 2013, "Impact of the Channel Thickness on the Performance of Ultrathin InGaAs Channel MOSFET Devices," *IEEE International Electron Devices Meeting*, Washington, DC, US, pp. 16.6.1–16.6.4.

Y. Bito, T. Kato and N. Iwata, 2001, *IEEE Transactions on Electron Devices*, vol. 48, p. 1503.

Edward-Yi Chang, Chien-I Kuo, Heng-Tung Hsu, Che-Yang Chiang, and Yasuyuki Miyamoto, 2013, "InAs Thin-Channel HighElectron-Mobility Transistors with Very High Current-Gain Cutoff Frequency for Emerging Submillimeter-Wave Applications," *Applied Physics Express*, vol. 6, no. 3, pp. 1–3.

Akira Endoh, Keisuke Shinohara, Issei Watanabe, Takashi Mimura, and Toshiaki Matsui, 2009, "Low-Voltage and High-Speed Operations of 30-nm-Gate Pseudomorphic $In_{0.52}Al_{0.48}As/In_{0.7}Ga_{0.3}As$ HEMTs Under Cryogenic Conditions," *IEEE Electron Device Letters*, vol. 30, no. 10, pp. 1024–1026.

Akira Endoh, Issei Watanabe, Keisuke Shinohara, Yuji Awano, Kohki Hikosaka, Toshiaki Matsui, Satoshi Hiyamizu, and Takashi Mimura, 2010, "Monte Carlo Simulations of Electron Transport in In0.52Al0.48As/In0.75Ga0.25As High Electron Mobility Transistors at 300 and 16K," *Japanese Journal of Applied Physics*, vol. 49, pp. 1–5.

Faiz Fatah, Chien-I Kuo, Heng-Tung Hsu, Che-Yang Chiang, Ching Yi Hsu, Yasuyuki Miyamoto, and Edward Yi Chang, 2012, "Bias Dependent Radio Frequency Performance for 40 nm InAs High Electron-Mobility Transistor with a Cut off Frequency Higher than 600 GHz," *Japanese Journal of Applied Physics*, vol. 51, pp. 1–3.

U. P. Gomes, K. Takhar, K. Ranjan, S. Rathi, and D. Biswas, 2012, "A Comparative TCAD Assessment of III-V Channel Materials for Future High Speed and Low Power Logic Applications," International Conference on Materials Science and Technology (ICMST 2012), IOP Publishing, vol. 73, pp. 1–4.

Ge Ji, Liu Hong-Gang, Su Yong-Bo, Cao Yu-Xiong, and Jin Zhi, 2012, "Physical Modeling Based on Hydrodynamic Simulation for the Design of InGaAs/InP Double Heterojunction Bipolar Transistors," *Chinese Physics B*, vol. 21, no. 5, pp. 1–6.

Dae-Hyun Kim, and Jesus A. del Alamo, 2008, "Lateral and Vertical Scaling of In0.7Ga0.3As HEMTs for Post-Si-CMOS Logic Applications," *IEEE Transactions on Electron Devices*, vol. 55, no. 10, pp. 2546–2553.

Tae-Woo Kim, Dae-Hyun Kim, and Jesus A. del Alamo, 2009, "30 nm In0.7Ga0.3As Inverted-Type HEMTs with Reduced Gate Leakage Current for Logic Applications," IEEE International Electron Devices Meeting, IEDM Technical Digest, pp. 1–4.

Tae-Woo Kim, Dae-Hyun Kim, and Jesus A. del Alamo, 2010, "60 nm Self-Aligned-Gate InGaAs HEMTs with Record High-Frequency Characteristics," *IEEE International Electron Devices Meeting*, pp. 30.7.1–30.7.4.

Tae-Woo Kim, and Jesus A. del Alamo, 2011, "Injection Velocity in Thin Channel InAs HEMTs," 23rd International Conference on Indium Phosphide and Related Materials, pp. 1–4.

P. H. Lai, S. I. Fu, Y. Y. Tsai, C. W. Hung, C. H. Yen, H. M. Chung, and W. C. Liu, 2006, *Journal of the Electrochemical Society*, vol. 153, p. G632.

J. W. Lee, S. J. Pearton, F. Ren, R. F. Kopf, J. M. Kuo, R. J. Shul, C. Constantine, and D. Johnson, 1998, *Journal of the Electrochemical Society*, vol. 145, p. 4036.

Wang Li-Dan, Ding Peng, Su Yong-Bo, Chen Jiao, Chan, Z.B, Zhang Bi-Chan, and Jin Zhi 2014,"100-nm T-gate InAlAs/InGaAsInP-based HEMTs with f_T= 249 GHz and f_{max} = 415 GHz," *Chinese Physics B*, vol. 23, no. 3, pp. 1–6.

Y. S. Lin, and Y. L. Hsieh, 2005, *Journal of the Electrochemical Society*, vol. 152, p. G778.

Y. S. Lin, D. H. Huang, W. C. Hsu, T. B. Wang, K. H. Su, J. C. Huang and C. H. Ho, 2006, *Semiconductor Science and Technology*, vol. 21, p. 540.

W. S. Lour, M. K. Tsai, K. Y. Lai, B. L. Chen and Y. J. Yang, 2001, *Semiconductor Science and Technology*, vol. 16, p. 191.

J. W. Matthews, 1975, *Journal of Vacuum Science and Technology*, vol. 12, p. 126.

A. Mohanbabu, N. Mohankumar, D. Godwin Raj, and P. Sarkar, 2017, "Investigation of Enhancement Mode HfO2 Insulated N-Polarity GaN/InN/GaN/In$_{0.9}$Al$_{0.1}$N Heterostructure MISHEMT for High-Frequency Applications," *Physica E*, vol. 92, pp. 23–29.

N. Mohankumar, Binit Syamal, and Chandan Kumar Sarkar, 2010, "Influence of Channel and Gate Engineering on the Analog and RF Performance of DG MOSFETs," *IEEE Transactions on Electron Devices*, vol. 57, no. 4, pp. 820–826.

Y. Okamoto, K. Matsunaga, and M. Kuzuhara, 1995, *Electronics Letters*, vol. 31, p. 2216.

D. Saguatti, A. Chini, G. Verzellesi, M. Mohamad Isa, K. W. Ian, and M. Missous, 2010, "TCAD Optimization of Field-Plated InAlAsInGaAs HEMTs," 22nd International Conference on Indium Phosphide and Related Materials (IPRM), pp. 1–3.

Hojjatollah Sarvari, Amir Hossein Ghayour, Zhi Chen, and Rahim Ghayour, 2016, "Analyses of Short Channel Effects of Single-Gate and Double-Gate Graphene Nanoribbon Field-Effect Transistors," *Journal of Materials*, pp. 1–8. Article ID 8242469, doi: https://doi.org/10.1155/2016/8242469.

E. F. Schubert, J. E. Cunningham, and W. T. Tsang, 1987, *Solid State Communications*, vol. 63, p. 591.

Suchismita Tewari, Abhijit Biswas, and Abhijit Mallik, 2013, "Impact of Different Barrier Layers and Indium Content of the Channel on the Analog Performance of InGaAs MOSFETs," *IEEE Transactions on Electron Devices*, vol. 60, no. 5, pp. 1584–1589.

Neha Verma, Mridula Gupta, R. S. Gupta, and Jyotika Jogi, 2013, "Quantum Modeling of Nanoscale Symmetric Double- Gate InAlAs/InGaAs/InP HEMT," *Journal of Semiconductor Technology and Science*, vol. 13, no. 4, pp. 342–354.

Niamh Waldron, Dae-Hyun Kim, and Jesus A. del Alamo, 2010, "A Self-Aligned InGaAs HEMT Architecture for Logic Applications," *IEEE Transactions on Electron Devices*, vol. 57, no. 1, pp. 297–304.

Y. Q. Wu, M. Xu, R. Wang, O. Koybasi, and P. D. Ye, 2009, "High-Performance Deep-Submicron Inversion-Mode InGaAs MOSFETs with maximum Gm exceeding 1.1 mS/µm: New HBr Pretreatment and Channel Engineering," *IEEE International Electron Devices Meeting (IEDM)*, pp. 1–4.

Fei Xue, Han Zhao, Yen-Ting Chen, Yanzhen Wang, Fei Zhou, and Jack C. Lee, 2011, "Effect of Indium Concentration on InGaAs Channel Metal-Oxide-Semiconductor Field-Effect Transistors with Atomic Layer Deposited Gate Dielectric," *Journal of Vacuum Science & Technology B, Nanotechnology and Microelectronics: Materials, Processing, Measurement, and Phenomena*, vol. 29, no. 4, pp. 1–4.

Zhong Yinghui, Wang Xiantai, Su Yongbo, Cao Yuxiong, Jin Zhi, Zhang Yuming, and Liu Xinyu, 2012, "Impact of the Lateral Width of the Gate Recess on the DC and RF Characteristics of InAlAs/InGaAs HEMTs," *Journal of Semiconductors*, vol. 33, no. 5, pp. 1–5.

Lisen Zhang, Zhihong Feng, Dong Xing, Shixiong Liang, Junlong Wang, Guodong Gu, Yuangang Wang, and Peng Xu, 2014, "70 nm Gate Length THzInP-Based In0.7Ga0.3As/In0.52Al0.48As HEMT With f$_{max}$ of 540 GHz," General Assembly and Scientific Symposium (URSI GASS), China, pp. 1–4.

4

Overview of THz Applications

T. Nagarjuna

CONTENTS

4.1 Introduction

Within the last generation, terahertz (THz) technology has developed rapidly, especially in spectroscopy and imaging applications. The main focus is on detecting THz frequency radiation for concealed weapons in nonpolar and nonmetallic materials, illicit drugs, and explosives. Moreover, this radiation does not pose a health risk for scanning. Plastic explosives and fertilizer bombs are weapons of war and support terrorism, and illicit drugs are a threat to society. The solution is to use the THz frequency radiations as electromagnetic (EM) waves to spectroscopically detect the materials through transmission characteristics in the range of 0.5–10 THz. This spectroscopy should be able to visualize the explosives or drugs even if they are concealed. Comparing this measured reflectivity THz spectra with known calibration spectra, people can quickly identify the presence of drugs, etc. (Federici et al. 2005).

This kind of radiation is invisible to the naked eye and is safe compared with that of x-ray radiation, and is used in pharmaceutical applications,

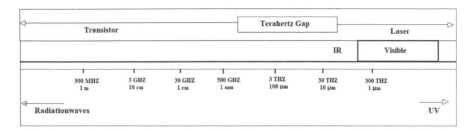

FIGURE 4.1
The spectroscopy of THz radiations with the frequency spectrum.

mainly in chemical analysis of tablets, capsules, and dosage forms. The spectroscopy of THz radiation is shown in Figure 4.1. The earth is a strong absorber of THz radiation in specific water vapor absorption bands. Generating this kind of radiation is challenging, but inexpensive commercial sources are present in the 300–1000 GHz range, such as gyrotron and tunnel diode (Pawar et al. 2013).

If the temperatures are more significant than 10 K, the THz radiations are emitted as part of blackbody radiation under natural phenomena. Artificial THz radiation sources include the gyrotron, backward wave oscillator, far-infrared quantum cascade, and free-electron lasers. The single-cycle sources used in time-domain spectroscopy are photoconductive and electro-optic emitters.

The resonant tunnel diode can be used as a source to generate the THz band at 542 GHz. In pharmaceutical care, THz radiation regulates ingredients in the human body. It covers the bioavailability of a drug as well as protects the stomach from high convergences of elements. The THz images are used to support the 3D investigation of tablets empowering the thickness of drugs and recognize the physical structure in medicine. The pharmacy industry mostly uses batch processing techniques, and if any contamination in the drug is identified, then the entire batch of drugs can be banned. Moreover, THz spectroscopy is useful for patenting pharmaceuticals.

4.2 Applications of THz Frequency

4.2.1 Dermatology

Terahertz frequency wavelength excites the intermolecular interactions, causing molecular vibration modes with spectroscopic information. Basal cell carcinoma is a common form of cancer identified worldwide in white populations. This carcinoma is reported in the USA for over 1 million people. The THz pulse imaging technique provides both structural and functional information of this type of cancer.

4.2.2 Oncology

It is estimated that 85% of all cancers take birth in the epithelium. Terahertz technology has the potential to improve conventional biopsy and associated surgery by more precise procedures. Utilizing THz radiation for oncology detection purposes will be more beneficial to society than the existing system.

4.2.3 Oral Health Care

With the help of THz radiation and imaging techniques, dental care problems can be solved with precise diagnostics. The imaging displays different types of tissues in the human tooth, and the early stage of diseases such as erosion of the tooth can be identified and treated better.

4.2.4 Medical Imaging

Terahertz radiation has low photon energy, thus does not damage tissues in the human body. This radiation can also detect the water content and density of tissues. It can penetrate through the skin into the epithelial layer without causing any harmful effects.

4.2.5 Security and Communication

Terahertz radiation can penetrate through fabrics and plastics used in surveillance, security screening, uncovering concealed weapons, fingerprint recognition, etc. It is mostly used in airports to detect metallic substances like ceramic knives and plastic explosives. The THz frequency serves for high-altitude telecommunications and aircraft-to-satellite communications.

Certain materials exist to have high absorbance peak positions at a higher frequency spectrum. Phonon modes mainly cause these peak positions because of the crystalline structure of these materials. This property can be utilized to detect drugs and explosives with THz scanning.

THz imaging has the potential to reveal concealed explosives in both metallic and nonmetallic weapons such as ceramic, plastic, knives, flammables, chemical weapons, and threats in packages. It even penetrates clothing and shoes. The THz radiation in the imaging modality provides identifiable representations with adequate standoff distance. It maintains penetration and resolution, as shown in Figure 4.2. The image should always be scanned stationary, and its speed should be in real time (Zimdars et al. 2004). The time-domain THz imaging determines three-dimensional structures with samples, and it gives multiple reflections from the optical system.

Millimeter waves operate in the range of 30 GHz, and they penetrate easily in barrier materials compared to THz waves. The essential factors that favor THz waves are spatial resolution and spectroscopic signatures.

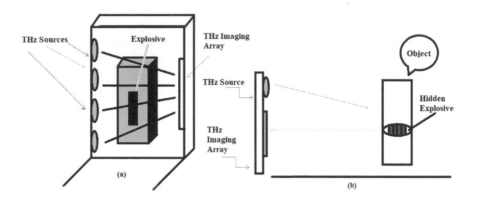

FIGURE 4.2
Penetration of THz radiation through fabrics and plastics used in surveillance, security screening, uncovering concealed weapons, and fingerprint recognition. (a) Transmission and (b) reflection modes of operation.

THz detection has roughly ten times better spatial resolution because their electromagnetic wavelength is ten times shorter than millimeter wave radiation.

4.3 Visualization Methods

The pulse variations depend on frequency; mostly, the parameters are absorption, optical thickness, and reflection. The superior signal-to-noise ratio (SNR) characteristics are observed very broadly in the time domain visualization method. These techniques represent the best detection methods for material classification (Loffler et al. 2002).

4.4 THz Generation in InAs HEMTs

4.4.1 Semiconductor Targets

An electro-optic emitter material widely used is zinc telluride (ZnTe). Similarly, for the photoconductive emitter, the material used is gallium arsenide (GaAs). Electro-optic material has a wide spectral range of 30 THz. The primary concern with this device is the bulk effect due to the increase in thickness and bandwidth variation. The photoconductive emitter generates radiation by creating photocarriers with bandgap illumination (Samoska et al. 2011).

4.4.2 P-Type InAs HEMTs

The cross section of the indium arsenide (InAs) device structure is as shown in Figure 4.3. This device design has an epitaxial layer to limit the parasitics and reduce the short-channel impacts. This device includes a 10 nm thick channel containing a 5 nm undoped InAs layer inside it. After a two-step recess process that exposes the indium aluminum arsenide (InAlAs) barrier, a Pt/Ti/Mo/Au (4/20/20/350 nm) gate is formed. The device is hardened at 250°C for 5 minutes to drive the platinum (Pt) into the InAlAs barrier.

Figure 4.4 shows the source injection velocity against gate length (L_g) for E-mode InAs pseudomorphic high electron mobility transistors (pHEMTs), as well as advanced Si CMOS, following the Antoniadis approach. InAs channel devices meet the ITRS requirement at L_g = 10 nm. Indium phosphide (InP) with high indium concentration HEMTs are an attraction for millimeter wave and THz applications. The outstanding transport properties of InAs channel material coupled with well-tempered design through insulator thickness scaling and buried Pt gate reduces the short-channel effects (Kim et al. 2008).

The first reported THz solid coordinated circuit amplifier was mainly dependent on a 25 nm InP HEMT process exhibiting amplification at 1 THz (1000 GHz) with 9 dB gain and an accessible increase at 1 and 1.5 THz anticipated f_{max} (Mei et al. 2015).

Generally, narrow bandgap semiconductors such as InAs exhibit high electron mobility. P-type InAs compound semiconductors have high efficiency,

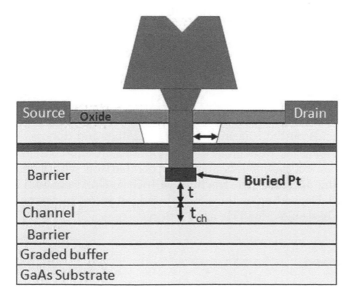

FIGURE 4.3
Cross-sectional view of the p-type InAs HEMT device with 5 nm InAs channel thickness and buried platinum material.

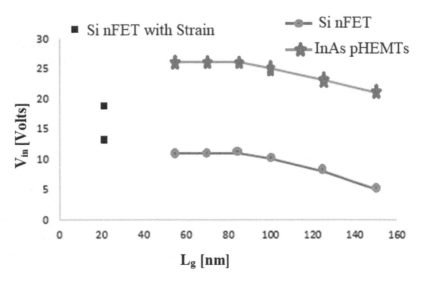

FIGURE 4.4

The subthreshold characteristics of source injection velocity vs. gate length for E-mode InAs pHEMTs, with InAs channel ITRS requirement at $L_g = 10$ nm.

TABLE 4.1

Physical Parameters of $In_{0.53}Ga_{0.47}As$, $In_{0.7}Ga_{0.3}As$, and InAs Materials

Parameter	Symbol	Unit	Material		
			$In_{0.53}Ga_{0.47}As$	$In_{0.7}Ga_{0.3}As$	InAs
Energy bandgap	E_g	E_v	0.74	0.58	0.36
Electron mobility	μ_e	cm^2/vs	11,400	11,000	20,000
Relative permittivity	ε_0	—	13,895	14.5	14.8
Electron peak drift velocity	—	cm/s	2.63×10^7	2.4×10^7	3.5×10^7
Electron affinity	X	e_V	4.72	4.665	4.9

which depends on the concentration of acceptors and donors. The parameters of the materials utilized in the simulations are shown in Table 4.1. The simulation model parameters and their comparison with transfer characteristics using different mole fractions of InGaAs/InAs channel HEMTs are investigated. Depending on the energy bandgap and electron mobility, these devices are best suited for THz applications.

4.5 Summary

In this chapter, an overview of THz applications and their effects are discussed. Various THz frequency-based applications like dermatology, oncology, oral health care, medical imaging, security, and communications are explored. The THz pulse imaging technique providing both structural and functional information is studied. The different visualization methods and their physical properties are discussed, along with their superior SNR characteristics.

References

Federici, John F., Brian Schulkin, Feng Huang, Dale Gary, Robert Barat, Filipe Oliveira, and David Zimdars. "THz imaging and sensing for security applications—explosives, weapons and drugs," *Semiconductor Science and Technology*, 20, no. 7 (2005): S266.

Kim, Dae-Hyun, and Jesus A. del Alamo. "30 nm E-mode InAs PHEMTs for THz and future logic applications," in *2008 IEEE International Electron Devices Meeting*, pp. 1–4, 2008.

Löffler, Torsten, Karsten Siebert, Stephanie Czasch, Tobias Bauer, and Hartmut G. Roskos. "Visualization and classification in biomedical terahertz pulsed imaging," *Physics in Medicine & Biology*, 47, no. 21 (2002): 3847.

Mei, Xiaobing, Wayne Yoshida, Mike Lange, Jane Lee, Joe Zhou, Po-Hsin Liu, Kevin Leong, et al. "First demonstration of amplification at 1 THz using 25-nm InP high electron mobility transistor process," *IEEE Electron Device Letters*, 36, no. 4 (2015): 327–329.

Pawar, Ashish Y., Deepak D. Sonawane, Kiran B. Erande, and Deelip V. Derle. "Terahertz technology and its applications," *Drug Invention Today*, 5, no. 2 (2013): 157–163.

Samoska, Lorene A. "An overview of solid-state integrated circuit amplifiers in the submillimeter-wave and THz regime," *IEEE Transactions on Terahertz Science and Technology*, 1, no. 1 (2011): 9–24.

Zimdars, David, and Jeffrey S. White. "Terahertz reflection imaging for package and personnel inspection in Terahertz for Military and Security Applications II," *International Society for Optics and Photonics*, 5411 (2004): 78–83.

5

Device and Simulation Framework of InAs HEMTs

V. Mahesh

CONTENTS

5.1 Introduction

One of the most efficient ways to increase the field-effect transistor performance to be used as a high-speed switching system is to reduce the device size. Recent advancements in manufacturing methods have made it possible to reduce the feature size to the sub-micrometer range. Channel electron transport can occur near ballistic levels and reach very high speeds in gallium arsenide (GaAs) devices where the channel length approaches the sub-micrometer regime. However, the reduction in size results in a smaller aspect ratio, leading to the degradation of specific system parameters, such as a significant change in threshold voltage, which is undesirable for developing high-quality integrated circuits (Ma et al. 2006).

When the operating frequency of the integrated circuit (IC) exceeds 100 GHz, the consideration of the device as a lumped parameter element becomes difficult. The current level is maintained constant, while its size, such as the high electron mobility transistor (HEMT) gate's width, is reduced in ICs. For intrinsic transconductance (g_m) improvement, a composite channel InGaAs or InGaAs/InAs providing high electron mobility and high saturation speed is preferable. In the past decade, the interest in developing new device architectures as InAs HEMTs in physics-based computer design has increased considerably. Standard techniques like drift–diffusion, however, are not adapted to nanoscale devices. They are unable to capture the strong electron confinement and the resulting energy quantization effectively. The approximation of the effective mass has recently been proposed for III–V devices to overcome these defects in quantum mechanical processes. However, the lowest drive band in the InAs is nonparabolic, so that the electron states cannot be adequately populated (Bass 1998).

Theoretically, the short-channel effect has two primary origins: the two-dimensional existence of the channel potential distribution and the space-charging current that flows under the channel into the substrate.

5.2 Short-Channel Effects

There are great challenges with significant scaling, in this case, the so-called short-channel effects (SCEs) in the form of unintended side effects. The transistors start to behave differently, which affects performance, model, and reliability if the MOSFET is of the same magnitude as the depleted layer width of the source and drain (Guerra et al. 2010). The major SCEs are

1. Drain-induced barrier lowering (DIBL)
2. Subthreshold slope (SS)

5.2.1 Drain-Induced Barrier Lowering (DIBL)

It is easier to understand this effect if we see the potential barrier profile to transfer from source to drain. There is a potential barrier under standard conditions ($V_{DS} = 0$ and $V_{GS} = 0$), which stops electron flow from source to drain (Fatah et al. 2015). As the gate voltage decreases, the potential barrier enables the electrons to flow. Ideally, the only voltage that influences the barrier will be this gate voltage. But, because the channel is narrower, a larger V_D extends the drainage area to reduce the potential barrier. The DIBL is given by

$$\text{DIBL} = \frac{\Delta V_T}{\Delta V_{ds}} \left[\frac{\left(V_{T1} - V_{T2} \right)}{\left(V_{ds1} - V_{ds2} \right)} \right] \tag{5.1}$$

Figure 5.1 clearly shows that the InAs HEMT with an optimized device structure produces reduced short-channel effects compared to conventional InGaAs HEMTs and Si FETs. Due to this superior property, this InAs-based HEMT achieves better device performance in IC design.

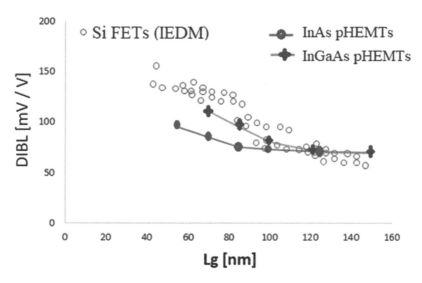

FIGURE 5.1
Variation of DIBL as a function of L_g for InAs HEMTs, InGaAs pHEMTs, and conventional Si FETs.

5.2.2 Subthreshold Slope

The subthreshold slope (SS) is a significant switching parameter that defines the ratio of the ON and OFF currents. To achieve a high ON/OFF current ratio with a low SS, the drain current has to be reduced by the same difference in V_{GS} by decades. Figure 5.2 clearly shows the variation of SS as a function of varying gate lengths for InAs HEMTs, InGaAs HEMTs, and conventional Si FETs. The figure shows that the InAs HEMT has a low SS value due to the optimized device structure. So these devices can be used for fast switching applications (Aizad et al. 2010).

5.3 Simulation Framework

5.3.1 Technology Computer-Aided Design (TCAD)

TCAD, or technology computer-aided design, is an electronic design automation tool that models device operation and IC fabrication. It uses computer simulations to describe the art of developing and optimizing the semiconductor processing technologies of the devices. The software involves different mathematical equations to model a self-designed device structure, structural properties, and electrical behavior. Sentaurus TCAD is one such software provided by Synopsys for the design and analysis of the devices.

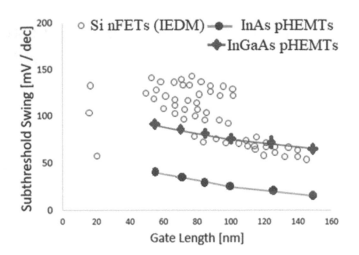

FIGURE 5.2
Variation of SS as a function of varying gate lengths for InAs HEMTs, InGaAs HEMTs, and conventional Si FETs.

TCAD tools are widely used today in the semiconductor industry and academic research. TCAD simulation and modeling can predict device performance, characterization, and explore the device development/optimization process for new technology. They can also help us to understand device physics and operation mechanisms.

Sentaurus Device is a numeric semiconductor device simulator capable of simulating various electrical, thermal, and optical characteristics of different semiconductor devices. It affects 1D, 2D, and 3D device behavior over a wide range of operating conditions, including mixed-mode circuit simulation, combining numerically simulated devices with compact modeling, performed on a SPICE-based circuit simulation level.

5.3.1.1 Input Files

Sentaurus Device expects at least one input file to define the device structure and the field values: mandatory doping-profile distributions and the optional mechanical-stress distribution inside a device.

An optional parameter file is also to be specified where material properties and physical model parameters are declared.

The grid file (in TDR format) includes the device geometry, the region and material specifications, contact and mesh definitions, the location of all the discrete mesh points, also called nodes or vertices, and the doping-profile distributions inside a device on a given mesh.

The grid file can represent 1D, 2D, or 3D device dimensions. It is the typical generation of the mesh engine Sentaurus Mesh. The file extension .tdr indicates that the file is in TDR format, which is the default format produced by Sentaurus Mesh.

The optional parameter file includes the specifications of the material parameters and user-defined model parameters. Parameter values specified in this file supersede the Sentaurus Device built-in defaults. The typical extension used for Sentaurus Device parameter files is .par.

5.3.1.2 Output Files

Sentaurus Device produces several output files:

- A file containing electrode names and resulting voltages, currents, charges, times, temperatures, and so on, which is indicated as the Current statement.
- A file with the spatially distributed solution variables and their derivatives, which is indicated as the Plot statement.
- A protocol file whose name is indicated as the Output statement.

For the Current file, Sentaurus Device always adds the _des.plt extension (if not specified explicitly) to the actual file name, such as nmos_des.plt.

The Plot specification indicates the file name. The final spatially distributed fields, such as the solution variables (carrier densities, electrostatic potential, and lattice temperature) and their derivatives, should be stored at the end of the simulations. The produced Plot file format depends on the layout of the input Grid file. If the Grid file is in TDR format, Sentaurus Device also has Plot output in TDR format. For the Plot file, Sentaurus Device always adds the extension _des to the actual file name, such as nmos_des.tdr.

The Output file specification instructs the Sentaurus Device where to accommodate the output generated during the device simulation. Sentaurus Device always adds the extension _des.log (if not specified explicitly) to the actual file name, such as nmos_des.log. The entire flow of input and output files is visualized in Figure 5.3.

The TCAD tool used for this work is Sentaurus, an advanced 1D, 2D, and 3D device simulator capable of simulating the electrical, optical, and thermal characteristics of semiconductor devices. We follow the path of a complete TCAD simulation project in device characterization, starting with device generation, device simulation, and finally analyzing the results obtained, as shown in Figure 5.4.

5.3.2 Device Simulation and Models

To optimize HEMTs, a reliable software simulation tool is required. Due to the presence of a high electric field in the device channel, a hydrodynamic approach is essential to model the electron transport over a two-dimensional cross section of the device.

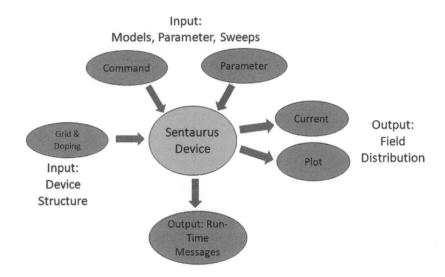

FIGURE 5.3
Flow chart of process simulation in Sentaurus TCAD to study the effects of short-channel effects for a 60 nm transistor. In each case, the simulation was carried out with changing parameters to visualize the electrical characteristics of the NMOS transistor.

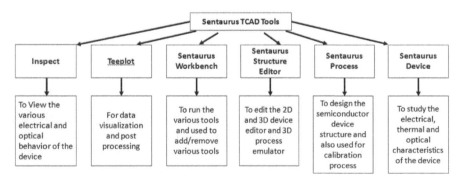

FIGURE 5.4
The complete flow of Sentaurus TCAD tools with 2D and 3D structure; device editor; structure generation using GUI; simulating the electrical, thermal, and optical characteristics of semiconductor devices; analyzing the data visualization; and processing of data. Inspection of different types of the electrical and optical behavior of the device can be accomplished.

5.3.2.1 Transport Models

Various transport models are available for simulation procedures such as thermodynamic, drift–diffusion, hydrodynamic, and Monte Carlo in the Sentaurus TCAD tool. Among these models, the user can incorporate any one of the models as per the requirement to analyze the performance of the device under investigation to get a high accuracy level in simulation. To specify the carrier transport in the semiconductor devices, the electron and holes continuity equation and Poisson equation are taken into account to accurately predict the carrier mobility in the channel.

The drift–diffusion (DD) carrier transport model in the TCAD tool considers carriers in the channel to be in thermal equilibrium with the lattice temperature. The hydrodynamic (HD) transport model in the simulator tool considers that the channel electron and hole temperature are not equal to the lattice temperature. So, the current density of the HD model expressions is more accurate than the DD model because of the influence of temperature gradients on the carriers. The simulation of devices through the HD model clearly shows that hot electrons dispersed in the barrier and the device channel layer are included. If the drain bias increases, the electrons get hotter, and the dispersion increases dramatically, leading to more traps in the buffer layer. It makes the HD model stand out from other models because of its high accuracy and fast compilation in nonequilibrium conditions.

The HD model from the 2D Sentaurus TCAD simulator tool is utilized to characterize the various device performances. The velocity overshoot in the depleted regions and quasi-ballistic transport in the device thin area are the nonequilibriums that can be predicted accurately using this HD model in the simulation. The device physical defects such as bandgap narrowing, effective variable mass, and doping-dependent mobility at high electric fields are considered. The carrier transport in high-speed devices such as heterostructure MOSFETs is in a nonequilibrium condition because of high energy electrons in the channel. Moreover, the electron velocity is very high at steady-state conditions.

The Poisson and continuity equations of the electron model are

$$\nabla \cdot \varepsilon \nabla \phi = -q(p - n + N_D - N_A) - \rho_{trap} \tag{5.2}$$

$$J_n = q\mu_n(n\nabla E_c + KT_n\nabla_n + Kn\nabla T_n - 1.5nkT_n\nabla \ln m_n) \tag{5.3}$$

where Φ, ε, n, p, q, ρ_{trap}, N_D, and N_A are the electrostatic potential, electrical permittivity, densities of electron and hole, electronic charge, fixed trap related to the charge density, and the concentration on ionized donors and acceptors, respectively. T_n is the electron temperature, m_n is the electron effective mass, and E_C is the conduction band energy. The equations from the transport model are solved at every meshing point in the device simulation.

The electron mobility model for the hydrodynamic simulation is to be specified by the following equations (Hemant Pardeshi et al. 2012):

$$\mu_{n,dop} = \mu_{min} + \frac{\mu_d}{1 + \left(\dfrac{N_{dop}}{N_0}\right)^A} \tag{5.4}$$

$$\mu_{n,field} = \frac{\mu_{no} + \left(\dfrac{V_{sat}}{E_0}\right)\left(\dfrac{F}{E_0}\right)^4}{1 + \left(\dfrac{F}{E_0}\right)^4} \tag{5.5}$$

$$\frac{1}{\mu_n} = \frac{1}{\mu_{n,dop}} = \frac{1}{\mu_{n,dop}} + \frac{1}{\mu_{n,field}} - \frac{1}{\mu_{no}} \tag{5.6}$$

where F is the driving force, and N_{dop} is the doping concentration. At the same time, μ_d and N_0 give the temperature-dependent coefficients, and A. μ_{no} is the low field mobility, $\mu_{n,\,dop}$ is the mobility due to impurity scattering, and field-dependent mobility at low doping concentration is given as $\mu_{n,\,field}$. The energy relaxation time (τ_E) of the materials used in this work is specified using the energy balance equation available in the HD model. The energy-dependent relaxation time for different combinations of materials utilized in this work is mentioned in Table 5.1. This HD model is efficiently rapid compared to the standard Monte Carlo transport model because the two-dimensional (2D) HD model is streamlined to compute only the average electron energy, so it is less time-consuming.

The Green's function method, which approximates the Shockley impedance field technique from the Sentaurus TCAD tool, is utilized to estimate the effect of gate-to-drain spectral densities. With this minimum noise figure (NF_{min}), noise resistance (R_n) and conductance (G_n) can also be measured. The carrier mobility in the channel is obtained using the Lombardi and Canali model from the TCAD simulation tool. The mobility of the carriers available in the device dual quantum well is observed by incorporating this model in the simulation procedure. The Shockley–Read–Hall (SRH) model is included in this simulation procedure to determine the riposte of the generation and recombination (G-R) of carriers in the channel. The carrier lifetime of holes

TABLE 5.1

Physical Parameters of the Materials Included in Device Simulation

Material	InAs	$In_{0.7}Ga_{0.3}As$	$In_{0.52}Al_{0.4}As$	InP
Energy bandgap, E_g (eV)	0.36	0.58	1.47	1.36
Conduction band offset, ΔE_c (eV)	\multicolumn{4}{c}{$In0.52Al0.48As/In0.7Ga0.3As = 0.62eV$; $In0.52Al0.48As = 3.1eV$}			
Electron mobility, μ_e (cm²/Vs)	20,000	11,000	4100	5400
Energy relaxation time, τ_e (ps)	0.08	0.17	0.1	0.5
Carrier lifetime, τ (ns)	$\tau_n = 1$	$\tau_n = 0.3$	$\tau_n = 100$	$\tau_n = 10$
	$\tau_p = 10$	$\tau_p = 10$	$\tau_p = 100$	$\tau_p = 3e3$
Relative permittivity, ε_0	14.8	14.5	12.46	12.6
Electron peak drift velocity (10^7 cm/s)	3.5	2.4	2.1	2.5
Lattice constant	\multicolumn{4}{c}{$In_xGa_{1-x}As = 5.6533 + 0.4051x$; $InAs = 6.0584$}			

(τ_p) and electrons (τ_n) of various materials in the proposed device architecture are included in this SRH model.

5.4 Summary

In this chapter, the fundamental physics of InAs-based HEMTs for improved device architectures is discussed. In the case of InAs-based HEMT devices, some basic rules are incorporated for device architecture and simulation procedures. Innovative CAD devices are broadly utilized today in the semiconductor business and scholastic examination. TCAD recreation and display can be used to foresee and investigate the device advancement/streamlining measure for innovation. They can likewise assist us with understanding device material science and activity systems.

References

F. Aizad, H.-T. Hsu, C.-I. Kuo, L.-H. Hsu, C.-Y. Wu, E. Yi, G.-W. Huang, S. Tsai, "Logic Performance of 40 nm InAs/In$_x$Ga$_{1-x}$As Composite Channel HEMTs," *Int. Conf. Enabling Sci. Nanotechnol.* (2010) 1–2.

R. Bass, "0.1 μm AlSb/InAs HEMTs with InAs Subchannel," *Electron. Lett.* 34, no. 1 (1998) 1525–1526.

F.A. Fatah, Y.-C. Lin, T.-Y. Lee, K.-C. Yang, R.-X. Liu, J.-R. Chan, H.-T. Hsu, Y. Miyamoto, E.Y. Chang, "Potential of Enhancement Mode In$_{0.65}$Ga$_{0.35}$As/InAs/In$_{0.65}$Ga$_{0.35}$As HEMTs for Using in High-Speed and Low-Power Logic Applications," *ECS J. Solid State Sci. Technol.* 4 (2015) N157–N159.

D. Guerra, R. Akis, F.A. Marino, D.K. Ferry, S.M. Goodnick, M. Saraniti, "Aspect Ratio Impact on RF and DC Performance of State-of-the-Art Short-Channel GaN and InGaAs HEMTs," *IEEE Electron Device Lett.* 31 (2010) 1217–1219.

B.Y. Ma, J. Bergman, P. Chen, J.B. Hacker, G. Sullivan, G. Nagy, B. Brar, "InAs/AlSb HEMT and Its Application to Ultra-Low-Power Wideband High-Gain Low-Noise Amplifiers," *IEEE Trans. Microw. Theor. Tech.* 54 (2006) 4448–4455.

6

Single-Gate (SG) InAs-based HEMT Architecture for THz Applications

M. Arun Kumar

CONTENTS

6.1 Introduction to Single-Gate HEMT Devices

For many years, high electron mobility transistors (HEMTs) have been attracting researchers for their high electron transport to support high-speed and low-power applications. An HEMT is a high-speed heterojunction device that uses two-dimensional hole/electron gas as the carriers in a quantum well. HEMT devices are replacing field-effect transistors (FETs) with

DOI: 10.1201/9781003093428-6

outstanding performance at high frequencies and improved power density. The basic structure of an HEMT consists of carriers having coulombic interaction with the ionized dopant atoms that increase mobility in the channel. High switching devices are necessary in the field of microwave communications and the radio frequency (RF) technology spectrum. In recent years, researchers have improved the performance of HEMTs based on the selection of material, layer thickness, and doping concentration of III–V semiconductors (Kim et al. 2010). This chapter explores the different types of single-gate HEMT structures, channel materials, electron transport properties, and performance. The fabrication technologies of different single-gate HEMTs are also discussed. Moreover, the device structure suitable for terahertz (THz) applications is realized.

6.2 Channel Materials

After silicon, the next successive material used for the fabrication of semiconductor devices is gallium arsenide (GaAs). It is a composite of gallium and arsenic elements from the III–V periodic table, acting as a direct bandgap semiconductor and a zinc blend crystal structure. GaAs is suitable for epitaxial growth on the substrate using III–V semiconductors such as indium gallium arsenide (InGaAs) and aluminum gallium arsenide (AlGaAs). InGaAs and AlGaAs are used to manufacture semiconductor devices such as infrared light-emitting diodes, microwave frequency integrated circuits, laser diodes, monolithic microwave integrated circuits, solar cells, and optical windows (Chang et al. 2013).

6.3 Electron Transport Properties of the Material

A thin indium arsenide (InAs) layer bonded into the InGaAs channel with AlGaAs as a barrier shows high electron transport and device performance. The mobility and electron velocity of these heterostructures can be increased by enhancing the thickness and the insertion of the InAs layer at 300 K. InAlAs/InGaAs is utilized for the fabrication of HEMT devices for high frequency (THz) applications.

6.4 InAlAs/InGaAs-Based HEMTs

6.4.1 Structure of the Device

A brief introduction to the general features of the InAlAs/InGaAs heterostructure fabrication process is discussed. The fabrication processes depict the design information and its specific performance characteristics and limitations (Endoh et al. 2012). The InAlAs/InGaAs structure with an InAs-inserted-channel HEMT device is illustrated in Figure 6.1.

1. Material growth is one of the most important preliminary steps for semiconductor devices. Fabrication of the heterostructure on Fe-doped semi-insulating (100) InP substrate was done by molecular beam epitaxy (MBE). The InGaAs and InAlAs layers are lattice-matched with the InP substrate.

2. An undoped InAlAs spacer layer is typically made with a thickness of 20 Å to improve the channel carrier confinement. It provides low leakage on the isolation floor. It also acts as a buffer layer that is sandwiched between the layers of InAlAs/InGaAs superlattices.

3. The bandgap of the InAlAs barrier layer is slightly more significant than the InP etch-stop layer. The Schottky barrier height is also much lower, leading to a high gate leakage current and low breakdown voltage. Furthermore, the n-doped InAlAs carrier supply layer is followed by a thickness of 150 Å grown on the surface to avoid this condition.

4. The electron density of the carrier supply layer is increased by adding an extra selective etching step to the gate recess process and exposing the undoped InAlAs gate layer as a high Schottky barrier with 200 Å thickness. The critical point in the region was etched for gate contact without damaging the other layers.

5. Further, the n-doped InAlAs and InGaAs cap layers are deposited with a thickness of 250 Å and 100 Å, respectively. The cap layers can

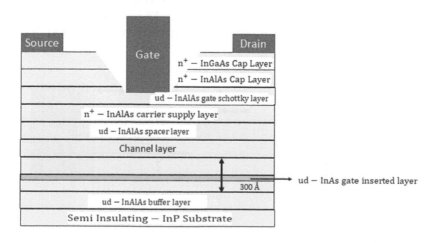

FIGURE 6.1
The structure of InAlAs/InGaAs with InAs-inserted-channel HEMT device formation of different layer thicknesses varying from 20 Å to 300 Å.

provide lower contact resistance and improve the uniformity process in gate recess etching. Because of its excellent etching selectiv-
ity, it is used to eliminate sidewall leakage current.

6. The selective areas and critical point regions are masked with a photoresist layer followed by the etching process to create gate, source,
and drain regions for metal contacts and finished with an intercon-
nection pattern. The critical growth of all layers is followed by a temperature of about 440°C since the layer thickness increases for a low
growth temperature.

Hence, the fabrication of InAlAs/InGaAs heterostructure has been deliber-
ated based on formation or deposition and patterning of different layers with
proper fabrication steps such as photoresist and the etching lift-off process.

6.4.2 Electrical Properties of InAlAs/InGaAs-Based HEMTs

The electron velocity for the InAlAs/InGaAs heterostructure with a thin InAs
layer implanted into the InGaAs channel was controlled using a beat (800 ns)
current–voltage estimation. Based on the structure, the velocity–electric field
(U-E) characteristics are measured. The device's top view structure looks like
an H-shaped bridge structure and is used to reduce the voltage drop at the
metal contacts shown in Figure 6.2. The bridge structures recess-etched area
with lengths (L) of 30, 40, 50, and 70 μm are as shown in Figure 6.2. The recess
etching can reduce the parallel conduction of the carrier supply layer and
the cap layer. The output current is measured for each recess-etched bridge
length by estimating the input voltage gradient for the four samples.

The Hall measurements provide the carrier density and velocity from the
output current. Hence, from the heterostructure's electrical measurement, it
is confirmed that the electron velocity of the InAs-inserted-channel heterostructure is higher than the conventional structure (Shinohara et al. 2002).

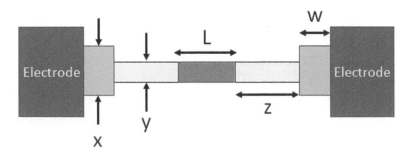

FIGURE 6.2
Electrical properties of InAlAs/InGaAs-based HEMTs. Schematic diagram of an H-shaped
design with the bridge structure recess-etched area of lengths (L) of 30, 40, 50, and 70 μm to
measure the electron velocity of the InAs-inserted-channel heterostructure.

6.5 Sub-50 nm Gate InP-Based HEMTs

6.5.1 Structure of the Device

The basic structure of sub-50 nm gate InP-based HEMTs (Figure 6.3) is explained in this section. In all other respects – masking, patterning, diffusion, photoresist, lithography, and lift-off process – the device's development is similar to InAlAs/InGaAs-based HEMT fabrication (Endoh et al. 2001). In summary, the typical processing steps are as follows:

- Mask 1—Defines the formation of epitaxial growth of lattice-matched InAlAs/InGaAs HEMT on semi-insulating (100) InP substrates by metalorganic chemical vapor deposition (MOCVD).
- Mask 2—Describes the layers from the bottom to the top design of the device; initially, a 300 nm InAlAs buffer layer is formed, then a 15 nm InGaAs channel layer is deposited.
- Mask 3—Used to place the spacer layer of InAlAs with 3 nm thickness, and it is followed by the deposition of the Si-δ-doped sheet in the range of 5×10^{12} cm^{-2}.
- Mask 4—Formation of a 6 nm InP etching stopper layer and 25 nm Si-doped n$^+$ InGaAs cap layer deposited in the range of 1×10^{19} cm^{-3} using chemical vapor deposition (CVD) at 250°C provides the uniformity in the gate recess etching and less contact resistance.
- Mask 5—The growth of all layers is carried out at 300°C to prevent contamination and reduce possible diffusion of the Si-δ-doped sheet.

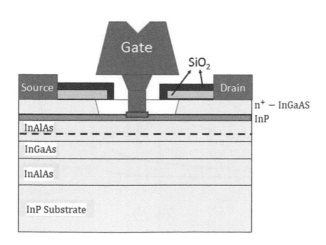

FIGURE 6.3
The structure of sub-50 nm gate InP-based HEMTs typical steps of Mask 1 to Mask 6 fabrication process of epitaxial growth of lattice-matched to ohmic contacts.

- Mask 6—Ohmic contacts are made to design the source and drain with 2 μm spacing and metals such as AuGe/Ni/Au at a 290°C substrate temperature in the presence of nitrogen for 10 minutes. The T-shaped gate is formed by Ti/Pt/Au with 50 × 2 μm using standard electron beam (EB) lithography and the lift-off technique.

6.5.2 I–V Characteristics of the 25 nm Gate HEMTs

The concept of the 25 nm gate HEMT involves the current–voltage characteristics from the gate-induced electric field in the channel between the source and drain as measured at room temperature, as shown in Figure 6.4. The arrangement of the layers show good pinch-off behavior. Meanwhile, the gate-to-source voltage (V_{gs}) induces the electron charge; drain current (I_{ds}) is dependent on both V_{gs} and the drain-to-source voltage (V_{ds}). The frequency obtained from the measurement is in the range of 396 GHz for the source–drain spacing with the length (L_{sd}) of 2 μm, drain–source voltage of 1 V, and

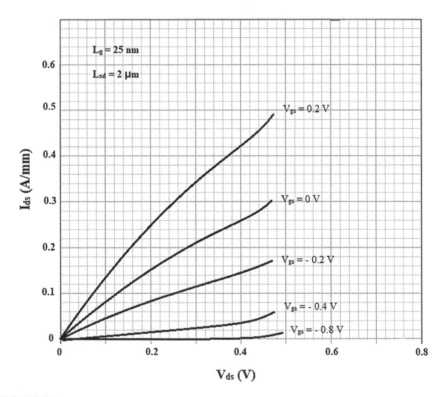

FIGURE 6.4
Current–voltage characteristics of 25 nm gate HEMT device. The gate-to-source voltage (V_{gs}) induces the electron charge, and the drain current (I_{ds}) is dependent on V_{gs} and V_{ds}.

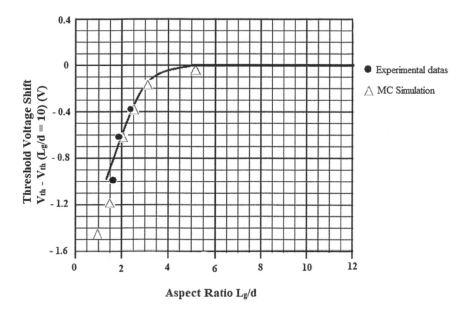

FIGURE 6.5
Channel aspect ratio versus threshold voltage shift. The threshold voltage shift is dependent on the threshold voltage and gate length.

gate–source voltage of −0.15 V. Another important aspect ratio is to avoid the short-channel effects from the critical channel based on the ratio L_g/d, where L_g is the length of the gate and d is the distance between the gate and drain shown in Figure 6.5. If the ratio is $L_g/d < 1$, threshold voltage shifts occur, confirming that short-channel effects are enhanced in the region. Hence, to increase the HEMT frequency, the channel aspect ratio (L_g/d) should be greater than 1.

6.6 Pseudomorphic $In_{0.52}Al_{0.48}As/In_{0.7}Ga_{0.3}As$-Based HEMTs

6.6.1 Structure of the Device

This is a brief introduction to the general aspects of HEMTs on III–V semiconductor materials relevant in the fabrication of InAlAs/InGaAs-based HEMTs used to design the pseudomorphic $In_xGa_{1-x}As/In_xAl_{1-x}As$ ternary alloy-based channel in HEMT devices.

The fabrication of devices based on pseudomorphic design enhances mobility and provides a high cutoff frequency of up to 500 GHz (Yamashita et al. 2002).

Before the fabrication process, it is important to discuss the general characteristics of the pseudomorphic $In_xGa_{1-x}As/In_xAl_{1-x}As$-based channel:

- The channel thickness for the material $In_xGa_{1-x}As$ should be employed in the range of $0.85 > x > 0.5$. The alloy scattering rate for the $In_xGa_{1-x}As$ material is directly proportional to $x(1-x)$. Due to the alloy scattering, the electron mobility increases with the value of x when $x < 0.5$ and decreases when $x > 0.5$.

- Generally, in OFF-state conditions, InGaAs small bandgap shows strong impact ionization, but it offers better breakdown characteristics and forward current–voltage curves in ON-state conditions. For the pseudomorphic channel, the situation is reversed: high mobility and large conduction occur at the band in offset, but the impact ionization will increase at ON-state for a small bandgap.

- The Schottky barrier layer is used to control the leakage current, and it is essential for developing semiconductor devices. The Schottky barrier is easy in normal devices but very difficult to grow Schottky contacts on the pseudomorphic channel. In the case of the pseudomorphic layers, they are very thin, hence, making the contacts in them is challenging. While fabricating, researchers should understand these basic layer characteristics and consider the channel distance to achieve high mobility in the device.

The fabrication process of pseudomorphic devices is similar in all aspects of normal HEMT fabrication. The typical approach for the structure of sub-25 nm gate InP-based pseudomorphic HEMTs is summarized as follows:

- Mask 1—Describes the epitaxial growth of pseudomorphic HEMTs on semi-insulating (100) InP substrates using MOCVD.

- Mask 2—The InAlAs buffer layer is formed 300 nm thick, then the $In_{0.7}Ga_{0.3}As$ channel layer is deposited.

- Mask 3—Grow the spacer layer of a 300 nm InAlAs and Si-δ-doped sheet.

- Mask 4—Deposit the 10 nm thick InAlAs barrier region and form the 6 nm InP etching stopper layer and 40 nm Si-doped n^+ $In_{0.53}Ga_{0.47}As$ cap layer deposited in the range of 1×10^{19} cm^{-3} using the CVD method.

- Mask 5—Diffusion of the Si-δ-doped sheet and ohmic contacts are made using the proper evaporated and lift-off process as the previous fabrication process.

6.6.2 I–V Characteristics of the Pseudomorphic Devices

The current–voltage characteristics of the sub-25 nm gate InP-based pseudomorphic HEMT measured at room temperature are shown in Figure 6.6.

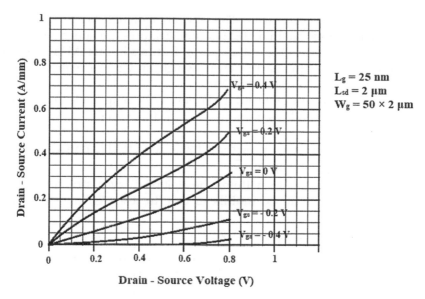

FIGURE 6.6
Current–voltage characteristics of the sub-25 nm gate InP-based pseudomorphic HEMT measured at room temperature.

The layers of the devices are completely pinched off and show good transconductance. The pseudomorphic device was designed similar to the InAlAs/InGaAs device with a T-shaped gate and the gate–drain voltage was measured at the standard channel aspect ratio ($L_g/d > 1$). Due to the reduced gate channel distance, the extrinsic time delay gets reduced and increases the electrons saturation velocity between the source and drain. Thus, the reduced gate-to-channel distance improves the cutoff frequency to 500 GHz for $In_{0.52}Al_{0.48}As/In_{0.7}Ga_{0.3}As$ ternary alloy-based HEMT devices, which is suitable for high-speed integrated circuits in optical communication systems as well as millimeter-wave and submillimeter-wave applications. If the gate thickness is increased to 40 nm, 50 nm, 60 nm, and 100 nm, it exhibits improved features in the cutoff frequency to drive high CMOS logic applications.

6.7 $In_{0.7}Ga_{0.3}As$-Based Metamorphic HEMTs

6.7.1 Structure of the Device

Research interests in the fabrication of HEMT devices originated in high-frequency communication systems. In particular, a device fabricated with the material growth of InAlAs/InGaAs-based metamorphic HEMTs (mHEMTs) on GaAs substrates provides outstanding speed performance of InP-based HEMTs with the low cost, increased robustness, and large wafer size offered by conventional GaAs processing. The fabrication of mHEMT devices is similar to HEMT devices, but in mHEMTs some features are added to improve the frequency response. One of the special features in mHEMTs, the metal (Pt) gate is buried in the barrier layer, which provides a realistic way to improve THz frequency (Kim et al. 2005).

The fabrication process of the mHEMT devices is comparable in all aspects of HEMT fabrication and is summarized in the following:

- Mask 1—The epitaxial growth of InGaAs mHEMTs on GaAs substrate, as shown in Figure 6.7.
- Mask 2—From top to bottom, the epitaxial layer structure consists of a heavily doped multilayer cap ($In_{0.7}Ga_{0.3}As/In_{0.53}Ga_{0.47}As/In_{0.52}Al_{0.48}As$):
  - 6 nm InP etch-stopper layer
  - 2 nm $In_{0.52}Al_{0.48}As$ barrier layer and upper Si δ-doping, 6 nm $In_{0.52}Al_{0.48}As$ barrier layer and lower Si δ-doping, 2 nm $In_{0.52}Al_{0.48}As$ spacer and 1 nm $In_{0.7}Al_{0.3}As$ spacer
  - 10 nm $In_{0.7}Ga_{0.3}As$ channel and 300 nm $In_{0.52}Al_{0.48}As$ buffer
- Mask 3—Deposit 0.3 μm categorized metamorphic buffer on the GaAs substrate. Here, dual Si δ-doping and an $In_{0.7}Al_{0.3}As$ spacer are used to lower the contact region's barrier potential and decrease the parasitic resistance.
- Mask 4—After the two-step recess process, the substrate exposes an InAlAs barrier. A Pt/Ti/Pt/Au gate is formed and annealed at 250°C for 2 minutes to drive the metal (Pt) into the InAlAs barrier.

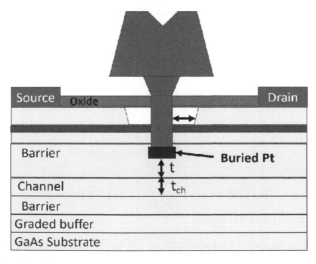

FIGURE 6.7
Schematic diagram of the metamorphic high electron mobility transistors (mHEMTs) on GaAs substrates.

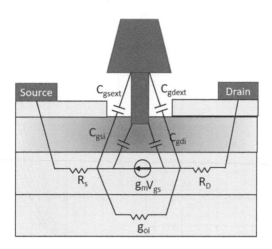

FIGURE 6.8
(a) The output characteristics of InGaAs mHEMTs on GaAs substrate. The V_{gs} induces the electron charge, and the I_{ds} is dependent on V_{gs} and V_{ds}. (b) The transconductance of InGaAs mHEMTs on GaAs substrate obtained with intrinsic parameters and extrinsic parameters.

6.7.2 DC Characteristics of $In_{0.7}Ga_{0.3}As$-Based MHEMTs

The device direct current (DC) characteristics are explored with the gate height of 250 nm and length (L_g) of 40 nm, and the side recess spacing (L_{side}) of 100 nm. The characteristics of two parameters – output and transconductance (g_m) – of the mHEMT device with L_g = 40 nm were measured to analyze the device gain as shown in Figure 6.8. The parameter transconductance describes the relationship between the drain–current (I_D) and input voltage (V_{gs}), whereas the output characteristics describe the good conductivity. An analytical model of the device is shown in Figure 6.9 with intrinsic parameters g_{mi}, g_{oi}, C_{gsi}, and C_{gdi}; and extrinsic parameters C_{gs-ext}, C_{gd-ext}, R_s, and R_d. The transit time (τ_t) is used to calculate the electron velocity under the gate. The parasitic charging delay (τ_{ext}) was reported through C_{gs_ext} and C_{gd_ext}. The parasitic delay (τ_{par}) is mainly associated with R_s and R_d. As L_g decreases, τ_t decreases consequently. However, both τ_{ext} and τ_{par} are not measured with L_g and remain constant. The electron velocity becomes 5×10^7 cm/s, which improves the cutoff frequency beyond 1 THz for microwave applications.

6.8 $In_{0.7}Ga_{0.3}As/InAs/In_{0.7}Ga_{0.3}As$-Based HEMTs

Several single-gate fabrication processes were discovered with InAlAs/InGaAs HEMTs, including $In_{0.7}Ga_{0.3}As/InAs/In_{0.7}Ga_{0.3}As$ composite-channel HEMTs using platinum (Pt) buried-gate technology, as shown in Figure 6.10. This HEMT device performance gained attention for its outstanding electronic transport properties and high saturation velocity (Kuo et al. 2008).

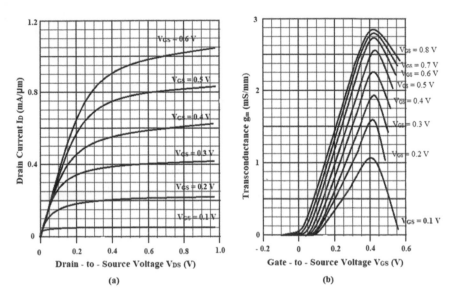

FIGURE 6.9
The analytical model of the device with intrinsic parameters g_{mi}, g_{oi}, C_{gsi}, and C_{gdi} and extrinsic parameters C_{gs-ext}, C_{gd-ext}, R_s, and R_d.

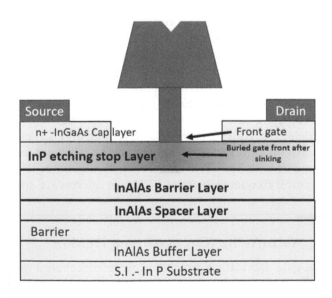

FIGURE 6.10
The schematic view of the $In_{0.7}Ga_{0.3}As/InAs/In_{0.7}Ga_{0.3}As$ composite channel HEMT device using buried gate.

The Pt buried gate technology was discussed earlier. Pt can diffuse into the barrier layer, and the channel can be further recessed to improve device performance. In this process, the InAs channel layer was developed with $In_{0.7}Ga_{0.3}As$ as the upper subchannel layer and $In_{0.7}Ga_{0.3}As$ as lower subchannel layers to enhance the electron confinement.

The remaining fabrication process is similar up to metal contacts. The device I–V characteristics with gate sinking technology show good pinch-off and saturation current due to their superior electron mobility. The performance of the device is attributed to an increase in transconductance and a decrease in capacitance.

6.9 InGaAs/Strained-InAs/InGaAs-Based HEMTs

$In_{0.52}Al_{0.48}As/In_xGa_{1-x}As$ is a promising device for future high-speed applications. In the previous section, the $In_{0.7}Ga_{0.3}As/InAs/In_{0.7}Ga_{0.3}As$ composite channel without considering strain in the InAs layer was discussed. The channel is modified with a strained band structure to develop the $In_{0.7}Ga_{0.3}As/strained-InAs/In_{0.7}Ga_{0.3}As$ composite HEMTs using Monte Carlo simulation, as shown in Figure 6.11. The composite channel has two-dimensional electron gas (2DEG), and Schrödinger and Poisson equations analyze it. The effect of 2DEG decreases the drain–source current (I_{ds}) but, the short-channel effects produce a negative threshold voltage shift to reduce the effects of 2DEG. The threshold voltage shift occurs in the region of $L_g/d < ~5$ and the cutoff frequency is improved up to 943 GHz without 2DEG and 813 GHz with 2DEG. The electron velocity in the region without 2DEG is

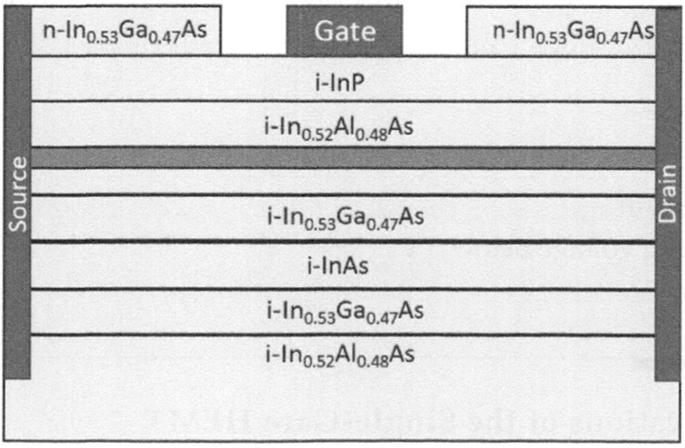

FIGURE 6.11
The cross-sectional view of the $In_{0.7}Ga_{0.3}As/strained-InAs/In_{0.7}Ga_{0.3}As$-based HEMT device for future high-speed applications.

higher than that with 2DEG. The channel electron velocity increases with a decrease in the channel length (L_g) in the simulation range. As a result, the drain current has been increased for the band structure $In_{0.7}Ga_{0.3}As$/strained-InAs/$In_{0.7}Ga_{0.3}As$ (Akazaki et al. 1992).

6.10 InAs Thin-Channel-Based HEMTs

There are several methods followed to develop different types of HEMT devices to achieve high frequency. They include gate-length scaling, reducing the barrier layer and thickness, reducing the parasitic resistances or capacitances, narrowing the source–drain spacing, and selecting the cap layer and spacer layer to enhance the electron transport properties. However, further scaling of the device dimensions has involved the InAs thin-channel HEMT and InAlAs barrier layer for future ultrahigh-speed applications. The heterostructure with a thin channel of InAs reduces the resistance across the Schottky barrier, thereby increasing the transconductance. The fabrication process of InAs thin-channel-based HEMTs is similar to normal HEMT devices. These thin-channel-based devices demonstrate outstanding DC and RF characteristics resulting in parasitic resistance/capacitance reduction with improved channel aspect ratio and output conductance.

6.11 Advantages of the Single-Gate HEMT

Major advantages of the InAs-inserted-channel heterostructure are as follows:

- Low noise
- Higher cutoff frequency
- Higher gain
- Operating voltage below 3 V

6.12 Applications of the Single-Gate HEMT

The main advantage of using an indium combination in the channel significantly improves the high-frequency operation of the HEMT device. Generally, these devices can be used in the RF domain as

- Highly efficient, high-power amplifiers for high-frequency applications
- High-speed optical infrared communication system integrated circuits (ICs)
- A passive switch for phased array applications

6.13 Summary

In this chapter, the fabrication concepts of single-gate InAs-based HEMT architecture were discussed along with different channel materials. Further, the I–V characteristics of various HEMT devices and their optimization to establish the THz frequency range are explored. The electrical properties of indium gallium arsenide (InGaAs), aluminum gallium arsenide (AlGaAs), and other heterostructures, including a thin InAs layer in the InGaAs channel, show high electron transport and device performance. The InAlAs/InGaAs distinct heterostructure can be fabricated as HEMTs for high-frequency (THz) applications. Herein, the structure of the devices and electrical properties of material like InAlAs/InGaAs-based HEMTs, sub-50 nm gate InP-based HEMTs, pseudomorphic $In_{0.52}Al_{0.48}As/In_{0.7}Ga_{0.3}As$-based HEMTs, $In_{0.7}Ga_{0.3}As$-based mHEMTs, and $In_{0.7}Ga_{0.3}As/InAs/In_{0.7}Ga_{0.3}As$-based HEMTs were discussed.

References

T. Akazaki, K. Arai, T. Enoki, Y. Ishii, "Improved InAlAs/InGaAs HEMT characteristics by inserting an InAs layer into the InGaAs channel," *IEEE Electron Device Lett.* 13 (1992) 325–327.

E.-Y. Chang, C.-I. Kuo, H.-T. Hsu, C.-Y. Chiang, Y. Miyamoto, "InAs thin-channel high-electron-mobility transistors with very high current-gain cutoff frequency for emerging submillimeter-wave applications," *Appl. Phys. Express.* 6 (2013) 34001.

A. Endoh, I. Watanabe, T. Mimura, "Monte Carlo simulation of InGaAs/strained-InAs/InGaAs channel HEMTs considering the self-consistent analysis of 2-dimensional electron gas," in *2012 Int. Conf. Indium Phosphide Relat. Mater.* (2012) 48–51.

A. Endoh, Y. Yamashita, K. Shinohara, M. Higashiwaki, K. Hikosaka, T. Mimura, S. Hiyamizu, T. Matsui, "Fabrication technology and device performance of sub-50-nm-gate InP-based HEMTs," in *Conf. Proceedings. 2001 Int. Conf. Indium Phosphide Relat. Mater. 13th IPRM (Cat. No.01CH37198)* (2001) 448–451.

C. Kim, J.A. del Alamo, "Scalability of sub-100 nm InAs HEMTs on InP substrate for future logic applications," *IEEE Trans. Electron Devices* 57 (2010) 1504–1511.

D.-H. Kim, J.A. del Alamo, J.-H. Lee, K.-S. Seo, "Performance evaluation of 50 nm $In_{0.7}Ga_{0.3}As$ HEMTs for beyond-CMOS logic applications," *IEEE Int. Devices Meet. 2005. IEDM Tech. Dig.* (2005) 767–770.

B. Kuo, H. Hsu, E.Y. Chang, C. Chang, Y. Miyamoto, S. Datta, M. Radosavljevic, G. Huang, C. Lee, "RF and logic performance improvement of $In_{0.53}Ga_{0.47}As/InAs/In_{0.53}Ga_{0.47}As$ composite-channel HEMT using gate-sinking technology," *IEEE Electron Device Lett.* 29 (2008) 290–293.

K. Shinohara, Y. Yamashita, A. Endoh, K. Hikosaka, T. Matsui, T. Mimura, S. Hiyamizu, "Extremely high-speed lattice-matched InGaAs/InAlAs high electron mobility transistors with 472 GHz cutoff frequency," *Jpn. J. Appl. Phys.* 41 (2002) L437–L439.

Y. Yamashita, A. Endoh, K. Shinohara, K. Hikosaka, T. Matsui, S. Hiyamizu, T. Mimura, "Pseudomorphic $In_{0.52}Al_{0.48}As/In_{0.7}Ga_{0.3}As$ HEMTs with an ultrahigh f/sub T/ of 562 GHz," *IEEE Electron Device Lett.* 23 (2002) 573–575.

7

Effect of Gate Scaling and Composite Channel in InAs HEMTs

C. Kamalanathan

CONTENTS

7.1 Introduction

High electron mobility transistors (HEMTs) based on indium phosphide (InP) have gained the attraction of the advanced wireless communication sector. They showed excellent performance at high frequencies due to their superior electronic transport properties and high saturation velocity. They are also potential candidates for low-power logic applications in a 22 nm node beyond silicon (Si) CMOS technology. For their excellent radio frequency (RF) performance with high current drivability, InP HEMTs usually utilize indium-rich InGaAs or InAs/InGaAs composite channels (Kruppa et al. 2006). The gate scaling structure also plays a vital role in the HEMT device's high-frequency performance. Generally, the transconductance (g_m)

DOI: 10.1201/9781003093428-7

of the device is mainly influenced by the distance between the gate and the channel. With optimized electron velocity under the gate electrode, the current–gain cutoff frequency (f_T) increases effectively. To further enhance the performance of InAs-based HEMTs, gate scaling techniques and composite channel techniques can be utilized while designing the device structure.

7.2 Gate Scaling

The development of gate scaling technology requires a successful manufacturing process of InAs/AlSb HEMTs. This step is crucial because the essential parameters for devices such as threshold voltage, power conduction, and gate-leakage current are defined. In other words, the scaling is crucial for the microwave and noise characteristics of HEMTs. The gate length is one of the critical parameters of HEMT device scaling. A reduction in the gate length is known to improve the RF characteristics such as cutoff frequency (f_T) and maximum oscillatory frequency (f_{max}). This is valid until the threshold voltage roll-off (V_{TH}), with reduced gate length as the aspect ratio of the gate to channel, is deteriorated. However, given that f_T and f_{max} depend upon several intrinsic and extrinsic parameters, consistent HEMT behavior with reduced gate length is preferred (Zhang et al. 2013).

7.3 Effect of Gate Scaling

7.3.1 Transfer Characteristics

In terms of design flexibility and integration, enhancement-mode HEMT (E-HEMT) devices offer many advantages over depletion-mode (D-HEMT) devices for circuit designers. The threshold voltage (V_{TH}) depends on the barrier layer thickness, permittivity of the barrier, discontinuity of the conduction band (ΔE_C) between the barrier and the channel layer, δ-doping density (N_d), and optimized Schottky barrier height as given in Equation 7.1.

$$V_{TH} = \left(\Phi_B - \Delta E_C - \frac{qN_\delta d_d^2}{2\varepsilon_d} \right) \tag{7.1}$$

where N_δ is the density of the silicon doping layers, d_d is defined as the distance between the channel and gate of the device, and the permittivity of the barrier layer is given by ε_d.

The value of $\Phi_B - \Delta E_C$ depends on channel layer materials, the barrier layer, and gate metal. The conduction band offset (ΔE_C) is directly proportional to the difference between energy bandgaps of the barrier and channel layer materials. Selecting a high work function gate metal and a thin barrier layer leads to the device enhancement mode operation with a positive threshold voltage (Suemitsu et al. 1999). The device with gate length L_g = 20 nm shows exceptional drain current saturation (I_{DSAT}) and pinch-off characteristics for V_{DS} = 0.8V. Figure 7.1a shows the variation of transconductance (g_m) and Figure 7.1b shows the drain current (I_D) as a function of the gate-to-source voltage at V_{DS} = 0.8V.

Figure 7.1 shows that for L_g = 20 nm and 40 nm, devices exhibit outstanding peak transconductance (g_m) of 3100 mS/mm and 2800 mS/mm, respectively, for a drain-to-source voltage (V_{DS}) of 0.8 V. Thus it is apparent that reducing the gate length will significantly boost the overall current driving capability of the device (Kim et al. 2006). The small gate-to-channel separation due to the inclusion of buried gate technology into the barrier layer minimizes the short-channel effects. It has the added advantage of low output conductance for small gate length devices (Guerra et al. 2010).

7.3.2 Output Characteristics

The output characteristics (I_D–V_{DS}) of the device for different gate voltages are shown in Figure 7.2. At V_{DS} = 0.8V, V_{GS} = 0.6V, and L_g = 20 nm and 40 nm, InAs-based HEMTs exhibit a drain current of 1360 mA/mm and 1140 mA/mm, respectively. This figure shows that a reduced gate length significantly improves the drain current at the same V_{GS} and V_{DS} (Harada et al. 1991).

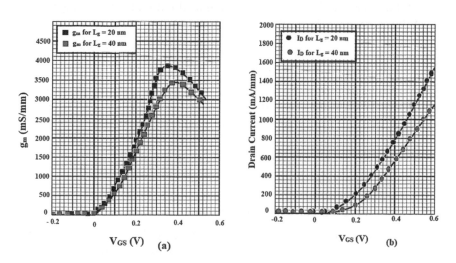

FIGURE 7.1
The variation of (a) transconductance (g_m) and (b) drain current (I_D) as a function of the gate-to-source voltage at V_{DS} = 0.8 V.

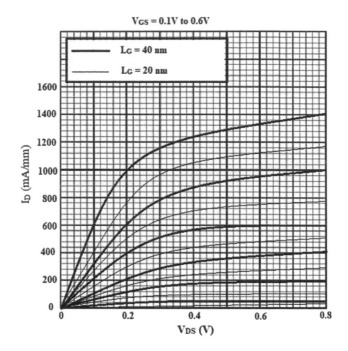

FIGURE 7.2
The output characteristics (I_D–V_{DS}) of the device for different gate voltages at V_{DS} = 0.8 V, V_{GS} = 0.6 V, and L_g = 20 nm, 40 nm.

Because of the aggressive gate scaling and other parameters, the device will exhibit very small ON resistance due to the dual Si–δ-doping on either side of the composite channel.

7.3.3 Short-Channel Effects

Scaling results in inhibiting the device performance due to the so-called short-channel effects (SCEs) in the form of unintended side effects that alter the device behavior (Yao et al. 2018). The major SCEs are

- Drain-induced barrier lowering (DIBL)
- Subthreshold slope (SS)

7.3.3.1 Drain-Induced Barrier Lowering (DIBL)

The gate length plays a crucial role in determining the short-channel effects of the device. The influence of gate length and channel thickness in determining the DIBL of the device is given in Figure 7.3. The figure shows that the devices with optimized gate length and reduced channel thickness result in reduced DIBL due to improved electrostatic integrity and scalability (Kharche et al. 2009).

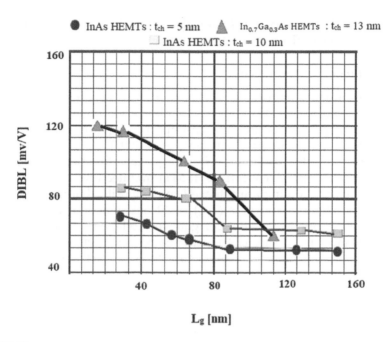

FIGURE 7.3
Variation of DIBL with gate length and channel thickness (T_{CH}).

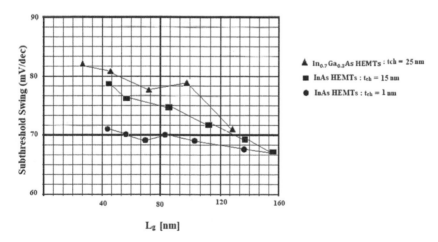

FIGURE 7.4
The subthreshold slope (SS) variation with gate length for different channel thicknesses (T_{CH}).

7.3.3.2 Subthreshold Slope (SS)

Figure 7.4 shows the subthreshold slope variation with gate length for different channel thicknesses (T_{CH}). Due to reduced electrostatic control, the subthreshold slope increases as the channel thickness increases. Moreover, the subthreshold slope decreases with gate length scaling. The minimum subthreshold slope observed is 83 mV/decade for L_g = 25 nm for the InAs-based HEMT device (Kim et al. 2008).

7.3.4 Frequency Performance

The RF performance metrics of the device are analyzed using two critical parameters, namely f_T and f_{max}. A high f_T plays a vital role in the design of high-speed digital circuits, and a much-improved f_{max} is desired for the efficient design of analog circuits (Matsuzaki et al. 2007).

Figure 7.5 shows the variation of f_T and f_{max} as a function of gate length for different dual-gate HEMT (DG-HEMT) architectures. High values of f_T and f_{max} with L_g scaling are achieved in composite channel DG-HEMTs with reduced channel thickness (T_{CH}) of 10 nm compared to other architectures. This is due to the superior gate controllability, reduced parasitic gate capacitance, and a shorter transit time of free electrons in composite channel DG-HEMTs than subchannel single-gate HEMTs (SG-HEMTs) and DG-HEMTs.

7.4 Effect of Composite Channel

The HEMT composite channel is introduced to advert the ionization of the InGaAs channel, one of the main constraints of InAs-based HEMTs. A significant element in achieving 300 GHz frequency performance is to optimize the indium composition in high-electron transmission channel structures on InAs (Kuo et al. 2008).

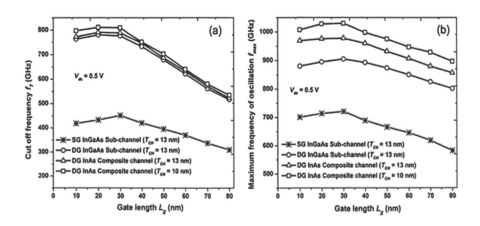

FIGURE 7.5
The variation of (a) f_T and (b) f_{max} as a function of gate length for different DG-HEMT architectures. High values of f_T and f_{max} with L_g scaling are achieved in composite channel DG-HEMTs with reduced channel thickness of 10 nm compared to other architectures.

7.4.1 DC Performance

The composite channel plays a crucial role in determining the electron density inside the quantum well of the device. The choice of materials and its conduction band energy are the determining factors in two-dimensional electron gas (2DEG) inside the well. The transfer characteristics (I_{ds}–V_{gs}) of InAs composite and InGaAs sub-channel devices for L_g = 30 nm and T_{CH} = 13 nm and 10 nm at V_{ds} = 0.5 V are given in Figure 7.6.

Due to the high electron density at the two accumulated charge regions created from the top and bottom gates, the drain current (I_D) of the DG-InGaAs HEMTs are higher than the single gate (SG) counterpart. The peak drain current (I_{ds}) for 13 nm and 10 nm DG-InAs composite channel devices are 1.2 and 1.14 mA/µm, respectively, compared to DG-InGaAs sub-channel HEMTs with I_{ds} of 0.97 mA/µm. The increased mobility of the InAs channel (µ at 20,000 cm²/Vs) compared to the InGaAs material (µ at 11,000 cm²/Vs) is crucial in determining the drain current.

The variation of transconductance as a function of gate voltage of InAs composite and InGaAs sub-channel devices of L_g = 30 nm and T_{CH} = 13 nm and 10 nm at V_{ds} = 0.5 V are shown in Figure 7.7.

The DG-InGaAs HEMT exhibits high transconductance (g_m) compared to the SG counterpart due to the presence of top and bottom delta doping layers with improved 2DEG and superior charge control. The results clearly

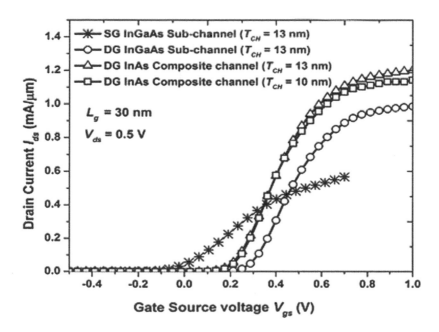

FIGURE 7.6
The transfer characteristics (I_{ds}–V_{gs}) of InAs composite and InGaAs sub-channel devices for L_g = 30 nm and T_{CH} = 13 nm, 10 nm at V_{ds} = 0.5 V.

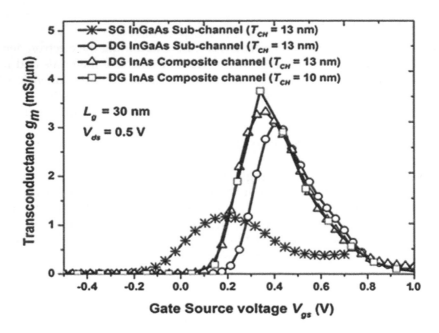

FIGURE 7.7
The variation of transconductance as a function of gate voltage for InAs composite and InGaAs sub-channel devices for L_g = 30 nm and T_{CH} = 13 nm, 10 nm at V_{ds} = 0.5 V.

show that the maximum g_m achieved for DG-InGaAs sub-channel HEMTs is 3.09 mS/μm for T_{CH} = 13 nm. Simultaneously, for DG-InAs composite channel HEMTs, the g_m is 3.3 and 3.7 mS/μm for T_{CH} = 13 nm and 10 nm, respectively at V_{ds} = 0.5 V (Radhakrishnan et al. 2018). The DG-InAs composite channel HEMT is compared with the InGaAs sub-channel counterpart for better carrier transport efficiency and electrostatic integrity representing higher transconductance.

7.4.2 Threshold Voltage (V_T)

The variation of threshold voltage (V_T) with the gate length (L_g) for InAs composite and InGaAs sub-channel devices for L_g = 30 nm and T_{CH} = 13 nm and 10 nm at V_{ds} = 0.5 V is shown in Figure 7.8. Here the DG-InAs composite and DG-InGaAs sub-channel achieve the V_T of 0.21 and 0.298 V at T_{CH} = 13 nm and L_g = 30 nm, and reach a maximum V_T of 0.27 V and 0.35 V at L_g = 80 nm, respectively at V_{ds} = 0.5 V compared with SG-HEMTs. Therefore, InAs composite channel DG-HEMTs, with a lower bandgap and high mobility, exhibit a lower positive V_T than DG-InGaAs. Improved gate scaling and a small channel facilitate enhancement mode operation in InAs composite channel DG-HEMTs that is much desirable for low-power applications.

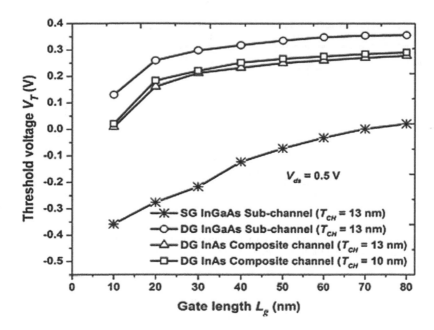

FIGURE 7.8
The variation of threshold voltage (V_T) with the gate length (L_g) for InAs composite and InGaAs sub-channel devices for L_g = 30 nm and T_{CH} = 13 nm, 10 nm at V_{ds} = 0.5 V.

7.4.3 RF Performance

The variation of cutoff frequency (f_T) and maximum oscillation frequency (f_{max}) with the gate-source voltage (V_{gs}) for InAs composite and InGaAs sub-channel devices for L_g = 30 nm and T_{CH} = 13 nm and 10 nm at V_{ds} = 0.5 V is shown in Figure 7.9. It is evident that the DG-HEMT architectures exhibit better RF efficiency with an inverted gate and a multicap structure. In contrast, excellent gate control causes quicker depletion and reduces the free electrons' transit time. The highest peak values of f_T/f_{max} of the DG-InGaAs sub-channel is 776/905 GHz. For DG-InAs composite channel devices f_T/f_{max} is 788/978 GHz for L_g = 30 nm at V_{ds} = 0.5 V.

7.5 Summary

In this chapter, a systematic investigation and comparison of InGaAs sub-channel and InAs composite channel DG-HEMT are performed by tuning the structural and geometrical parameters. The results are also compared with existing SG-HEMTs to show the necessity of DG-HEMTs in composite

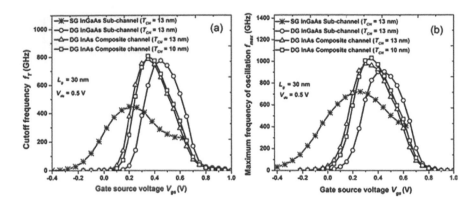

FIGURE 7.9

The variation of (a) cutoff frequency (f_T) and (b) maximum oscillation frequency (f_{max}) with the gate-source voltage (V_{gs}) for InAs composite and InGaAs sub-channel devices for L_g = 30 nm and T_{CH} = 13 nm, 10 nm at V_{ds} = 0.5 V.

and sub-channel devices. Moreover, InGaAs sub-channel (T_{CH} = 13 nm)/ InAs composite channel (T_{CH} = 13 nm, 10 nm) channel DG-HEMT devices exhibit a maximum g_m of 3.09/3.3/3.7 mS/µm, peak f_T of 776/788/809 GHz, and f_{max} of 905/978/1030 GHz at 30 nm gate length at V_{ds} = 0.5 V, respectively. Compared to the DG-InGaAs sub-channel, the thin DG-InAs composite channel shows improved DC, g_m, peak average electron velocity, and improved analog/RF performance for future terahertz and sub-millimeter wave applications.

References

D. Guerra, R. Akis, F.A. Marino, D.K. Ferry, S.M. Goodnick, M. Saraniti, "Aspect Ratio Impact on RF and DC Performance of State-of-the-Art Short-Channel GaN and InGaAs HEMTs," *IEEE Electron Device Lett.* 31 (2010) 1217–1219.

N. Harada, S. Kuroda, T. Katakami, K. Hikosaka, T. Mimura, M. Abe. "Pt-Based Gate Enhancement-Mode InAlAs/InGaAs HEMT's for Large-Scale Integration," *3rd Indium Phosphide and Related Materials Conference. IPRM* (1991) 377–80.

N. Kharche, G. Klimeck, D. Kim, J.A. del Alamo, M. Luisier, "Performance Analysis of Ultra-Scaled InAs HEMTs," *2009 IEEE Int. Electron Devices Meet.* (2009) 1–4.

D. Kim, J.A. del Alamo, "30 nm E-mode InAs PHEMTs for THz and Future Logic Applications," *2008 IEEE Int. Electron Devices Meet.* (2008) 1–4.

D.-H. Kim, J.A. Del Alamo, J.-H. Lee, K.-S. Seo, "The Impact of Side-Recess Spacing on the Logic Performance of 50 nm InGaAs HEMTs," *International Conference on Indium Phosphide and Related Materials Conference Proceedings, 2006* (2006) 177–80.

W. Kruppa, J. B. Boos, B. R. Bennett, N. A. Papanicolaou, D. Park, R. Bass, "InAs HEMT Narrowband Amplifier with Ultra-Low Power Dissipation," *Electronics Lett.*, 42, no. 12 (8 June 2006) 688–690.

C. Kuo, H. Hsu, E.Y. Chang, C. Chang, Y. Miyamoto, S. Datta, M. Radosavljevic, G. Huang, C. Lee, "RF and Logic Performance Improvement of $In_{0.7}Ga_{0.3}As/InAs/In_{0.7}Ga_{0.3}As$ Composite-Channel HEMT Using Gate-Sinking Technology," *IEEE Electron Device Lett.* 29 (2008) 290–293.

H. Matsuzaki, T.T. Maruyama, T. Koasugi, H. Takahashi, M. Tokumitsu, T. Enoki, "Lateral Scale Down of InGaAs/InAs Composite-Channel HEMTs with Tungsten-Based Tiered Ohmic Structure for 2-S/mm g_m and 500-GHz f_T," *IEEE Trans. Electron Devices* 54 (2007) 378–384.

S.K. Radhakrishnan, B. Subramaniyan, M. Anandan, M. Nagarajan, "Comparative Assessment of InGaAs sub-Channel and InAs Composite Channel Double Gate (DG)-HEMT for Sub-Millimeter Wave Applications," *AEU - Int. J. Electron. Commun.* 83 (2018) 462–469.

T. Suemitsu, H. Yokoyama, Y. Umeda, T. Enoki, Y. Ishii. "High-Performance 0.1- μm Gate Enhancement-Mode InAlAs/InGaAs HEMTs Using Two-Step Recessed Gate Technology," *IEEE Trans. Electron. Dev.* 46 (1999) 1074–80.

J. Yao, Y. Lin, H. Hsu, K. Yang, H. Hsu, S.M. Sze, E.Y. Chang, "Evaluation of a 100-nm Gate Length E-Mode InAs High Electron Mobility Transistor with Ti/Pt/Au Ohmic Contacts and Mesa Sidewall Channel Etch for High-Speed and Low-Power Logic Applications," *IEEE J. Electron Devices Soc.* 6 (2018) 797–802.

X.F. Zhang, L. Wei, L. Wang, J. Liu, J. Xu, "Gate Length Related Transfer Characteristics of GaN-Based High Electron Mobility Transistors," *Appl. Phys. Lett.* 102 (2013) 113501.

8

Double-Gate (DG) InAs-based HEMT Architecture for THz Applications

R. Poornachandran

CONTENTS

8.1 Introduction

Conventional silicon complementary metal–oxide–semiconductor (CMOS) device scaling has reached the end of the road map following Moore's law (Li et al. 2012). However, the short-channel effect, high power dissipation, and low thermal stability constraints degrade device performance. High electron mobility transistors (HEMTs) with compound III–V semiconductors play a vital role in the present CMOS era in millimeter wave and sub-millimeter wave applications because they can work with a low noise margin on the terahertz scale. These devices have high carrier mobility characteristics, saturation speed, and higher channel charge densities than conventional silicon (Si)-based CMOS devices (Endoh et al. 2009). The direct current (DC) and radio frequency (RF) characteristics are increased in HEMTs through multilayer cap technology, reducing the barrier and channel thickness, and reducing ON-resistance through a high stem height recessed gate structure.

A high cutoff frequency (f_T) value and a maximum oscillation frequency (f_{max}) above 700 GHz are required for future communications systems in the recent literature. For a better establishment of analog circuits, high f_T plays a crucial role in designing high-speed digital circuits and increases f_{max} (Ajayan et al. 2017).

Aggressive device scaling boosts the RF efficiency and increases the short-channel effects (SCEs) as well. Simultaneously, the efficacy of the regulation of gates is also decreased as two-dimensional electron gas (2DEG) electrons are weakly sealed on the channel. Double-gate HEMTs (DG-HEMTs) with recessed technology design, having two gate electrodes on each side of the conducting channel up to the InP etch stop layer, will mitigate such effects. The DG structure improves gate modulation efficiency in the conduit with reduced SCEs and the device high analog/RF capacity (Yu et al. 2013).

8.2 Device Structure

Figure 8.1 shows the structure of DG-HEMTs on the InP substrate with a composite channel. In this structure, multicap layers of 5 nm $In_{0.65}Ga_{0.35}As$

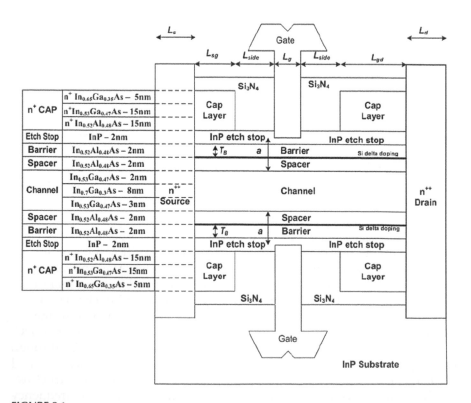

FIGURE 8.1
The structure of the composite channel double-gate HEMT on the indium phosphate substrate. Multiple layers are present in the design structure.

(2×10^{19} cm^{-3}), 15 nm In$_{0.53}$Ga$_{0.47}$As (2×10^{19} cm^{-3}), and 15 nm In$_{0.52}$Al$_{0.48}$As (2×10^{18} cm^{-3}) are introduced in the device structure to decrease the contact and source/drain parasitic resistance (Kumar et al. 2018). A 2 nm InP etch stop layer is used below the cap layer. Next, an In$_{0.52}$Al$_{0.48}$As barrier layer is inserted, followed by a thin In$_{0.52}$Al$_{0.48}$As Si-delta doping layer (1×10^{18} cm^{-3}) supplying excess carriers to the channel. An undoped 2 nm thin spacer layer of In$_{0.52}$Al$_{0.48}$As is placed between the barrier and the channel layer to overcome the effect of alloy and impurity scattering. The device contains a subchannel that includes a thin 8 nm In$_{0.7}$Ga$_{0.3}$As sub-channel layer with 2 nm/ 3 nm In$_{0.53}$Ga$_{0.47}$As upper/lower channel layers to enhance carrier transport properties by improving analog/RF device performance.

The gate is embedded in the etch stop layer to increase the gate control to the channel, and gate length (L_g) is optimized and fixed at 50 nm. The source gate ($L_{sg} + L_{side}$) and drain gate ($L_{side} + L_{gd}$) spacing are 0.5 and 1 µm, respectively (Takahashi et al. 2017). Source and drain (S/D) contacts are extended to the spacer layers so that the electrons can be moved to the channel immediately.

Highly doped In$_{0.6}$Ga$_{0.4}$As is used as a source and drain with an active arsenic concentration of 1×10^{20} cm^3 with ohmic contacts. The top and bottom Schottky gate is made up of gold (Au) to control the channel. All of the above active layers are symmetrical with respect to the channel forming the DG structure. These DG-HEMTs operate under the symmetric condition in which the voltages applied to both gate terminals are equal.

8.3 Device Performance

8.3.1 DC Performance

Figure 8.2a shows the variation of the drain current (I_{ds}) and transconductance (g_m) for the gate–source voltage (V_{gs}) of DG-HEMTs. This is done for different L_g (30, 60, 150 nm) and barrier thickness (T_B) 2 nm at $V_{ds} = 0.5$ V. Figure 8.2b shows the variation of g_m with different L_g for $T_B = 2$ nm at $V_{ds} = 0.5$ V. From Figure 8.2a, it is clear that the peak value of the drain current I_{ds} and g_m obtained was 0.98 mA/µm and 3.09 mS/µm at $L_g = 30$ nm and $T_B = 2$ nm.

The DG control of the conductive channel in the DG-HEMT device with reduced gate-to-channel distance was responsible for faster depletion with improved electron density. Improving the mean drift velocity in the channel under the gate electrodes control for thin T_B results in higher peak I_{ds} and g_m with In$_{0.53}$Ga$_{0.47}$As/In$_{0.7}$Ga$_{0.3}$As/In$_{0.53}$Ga$_{0.47}$As DG-HEMT devices and is found to be much higher than its single-gate (SG) counterpart (Pati et al. 2013).

The output characteristic I_{ds}–V_{ds} of the DG-HEMT is displayed in Figure 8.3 for values of L_g 30, 60, and 150 nm for different V_{gs}. From this graph, it is

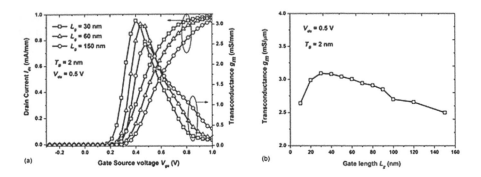

FIGURE 8.2

(a) Drain current and transconductance variation with the change of gate–source voltage. The graph was plotted with different gate lengths of 30 nm, 60 nm, and 150 nm; barrier thickness of 2 nm; and drain–source voltage of 0.5 V. It shows that the drain current has a firm value concerning the gate–source voltage. (b) The transconductance for the device's gate length. The transconductance increased initially up to 30 nm in length. After that, it showed a slow decrease in transconductance.

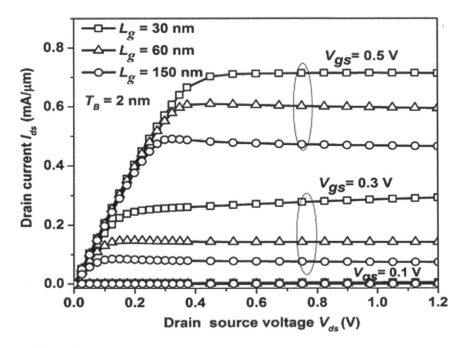

FIGURE 8.3

The plot of double-gate HEMT output characteristics. At the lower drain–source voltage, the drain current increased for a particular gate–source voltage. As the length decreases, the drain current shows a positive response. Once it reaches a saturation condition, the current becomes constant.

evident that as the L_g decreases, the I_{ds} increases for fixed V_{gs}. Maximum saturated I_{ds} is 0.7 mA/μm for $L_g = 30$ nm and $T_B = 2$ nm for $V_{gs} = 0.5$ V. In the proposed structure, increased gate control in DG-HEMTs leads to a high density of 2DEG sheet carrier with lower SCEs owing to the higher $I_{ds,sat}$ values (Kuo et al. 2009).

8.3.2 Threshold Voltage (V_T)

The threshold voltage difference (V_T) is shown in Figure 8.4 for DG-HEMTs of different L_g and T_B. A high positive V_T of 0.2 to 0.398 V was achieved with $T_B = 1$ nm and $V_T = 0.13$ V to 0.374 V for $T_B = 2$ nm and $L_g = 10$ nm to 150 nm is observed. It is noted that, because of an increase in parasite resistance, the V_T is changing more positively with L_g values. When T_B increases, V_T decreases due to the increased gate-to-channel distance due to the reduction in electro-static gate control (Jaiswal et al. 2015).

8.3.3 Short-Channel Effects

One of the significant issues with modern HEMTs is their inherently reduced electrostatic gate power, as the 2DEG channel is located away from the gate.

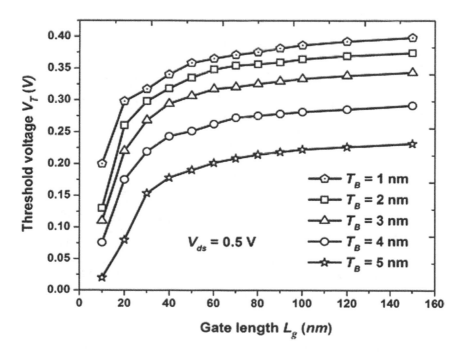

FIGURE 8.4
The threshold voltage variation for the gate length holding a constant barrier thickness. As the thickness increased for double-gate HEMT, the threshold voltage noticeably decreased.

The dominant influences of SCEs are drain-induced barrier lowering (DIBL) and subthreshold slope (SS) (Park et al. 2011).

8.3.3.1 Subthreshold Slope (SS)

Figure 8.5 shows SS dependence on L_g with varying values of T_B. SS comes from the I_{ds}–V_{gs} characteristic curve slope at $V_{ds} = 0.5$ V. SS is known as the voltage variation in gate voltage needed for the subthreshold drain curve to produce one decade changes. A thinner T_B device provides excellent electrostatic control, which results in lower SS. This effectively reduces leakage current in the OFF-state. As T_B increases, electrostatic control will cease to increase the SS. A linear increase in SS from 77.6 mV/decade for $T_B = 1$ nm, 81 mV/decade for $T_B = 2$ nm to 95.2 mV/decade for $T_B = 5$ nm with $L_g = 10$ nm is observed.

8.3.3.2 Drain-Induced Barrier Lowering (DIBL)

The essential parameter that describes the electrostatic integrity of HEMTs is DIBL, which is defined as the change of V_T to V_{ds} (DIBL $= \Delta V_T / \Delta V_{ds}$). As

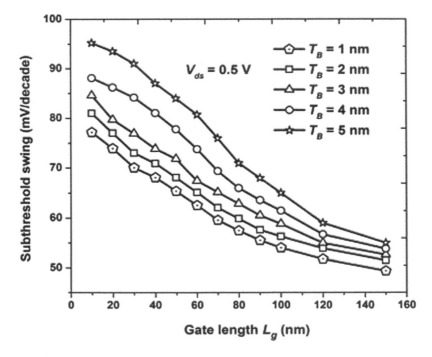

FIGURE 8.5
The pictorial view of the subthreshold slope for the gate length variation. With a lower gate length, the subthreshold slope is more compared to a longer length. The barrier thickness also has a prominent effect on this. Less thickness leads to low subthreshold swing. The graph was plotted with a drain–source voltage of 0.5 V.

the T_B becomes thicker, the channel moves away from the gate, reducing the control of the electrostatic gate. Reduction in control of the gate leads to the degradation of the DIBL. There is a linear increase in DIBL from 68 mV/V for T_B = 2 nm to 82 mV/V for T_B = 5 nm for L_g = 30 nm. A decrease in DIBL with an increase in L_g is observed from Figure 8.6. A decrease in L_g increases the lateral electrical field, penetrating from the source to the drain, thereby increasing the DIBL. The SCE is highly noticeable when the L_g is less than 30 nm.

8.3.4 RF Performance

The device RF performance is calculated with two important parameters, f_T and f_{max}. f_T is the frequency at which the current gain is unity, and f_{max} is the frequency at which the power gain is unity (Royter et al. 2003). Figure 8.7a and Figure 8.7b show variations in f_T and f_{max} with respect to V_{gs} for DG-HEMT devices with different T_B values for L_g = 30 nm. In DG-HEMTs, the f_T and f_{max} obtained are higher than in SG-HEMTs due to higher g_m and I_{ON}. The highest f_T and f_{max} values observed are 776 and 905 GHz (T_B = 2 nm), and 804 and 920 GHz (T_B = 1 nm), respectively, for L_g = 30 nm and V_{ds} = 0.5 V.

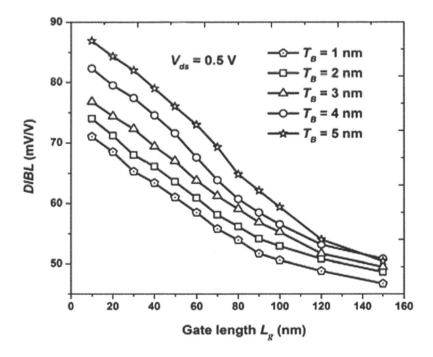

FIGURE 8.6
The plot shows the variation of drain-induced barrier lowering concerning gate length change. The drain-source voltage was kept at a fixed 0.5 V. The barrier thickness plays a prominent role here. An increase in thickness barrier increases the DIBL.

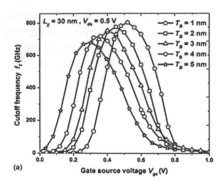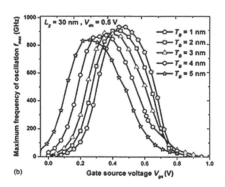

FIGURE 8.7

(a) The variation of cutoff frequency with the change of gate–source voltages. The cutoff voltage increased exponentially; once it reached the peak value, it slowly started decreasing again and returned to the initial position. When the thickness increased, the cutoff frequency got more weight at the later stage of the gate–source voltage. (b) The maximum frequency of oscillation concerning the gate–source voltage. It appears similar to the cutoff frequency plot.

8.4 Summary

Throughout this chapter, the influence of L_g and T_B on DG-HEMTs with subchannels $In_{0.53}Ga_{0.47}As/I_{0.7}Ga_{0.3}As/In_{0.53}Ga_{0.47}As$ is explained in detail. Because of optimized DIBL and SS, the higher values of I_d and g_m are achieved along with the better performance with positive V_T. For T_B thinning, the frequency efficiency is improved by high g_m and low parasitics. As T_B is less than 2 nm, the leakage current rises and the I_{ON} decreases simultaneously. On the other hand, if L_g is smaller than 30 nm, SCEs become effective, thus degrading the output. In comparison with conventional InGaAs HEMTs, the proposed $In_{0.53}Ga_{0.47}As/In_{0.7}Ga_{0.3}As/In_{0.53}Ga_{0.47}As$ optimized DG-HEMT device exhibits a maximum g_m of 3.09 mS/µm, $V_T = 0.29$ V, I_{ON}/I_{OFF} current ratio of 2.24 × 10⁵, SS ~73 mV/decade, DIBL ~68mV/V, and maximum f_T and f_{max} of 776 GHz and 905 GHz for $L_g = 30$ nm and $T_B = 2$ nm at $V_{ds} = 0.5$ V. These aspects make it suitable for high-frequency and high-speed applications.

References

J. Ajayan, D. Nirmal, "20-nm Enhancement-Mode Metamorphic GaAs HEMT with Highly Doped InGaAs Source/Drain Regions for High-Frequency Applications," *Int. J. Electron.* 104 (2017) 504–512.

N. Chavan Jaiswal, D. Kulshrestha, H. Pardeshi, "Effect of Gate Length Variation on DC Performance of Ino.7Gao.3As/InAs/Ino.7Gao.3As Composite Channel HEMT", *2015 IEEE International Conference on Computer Graphics, Vision and Information Security (CGVIS), Bhubaneswar* (2015) 181–184.

A. Endoh, K. Shinohara, I. Watanabe, T. Mimura, T. Matsui, "Low Voltage and High-Speed Operations of 30-nm-Gate Pseudomorphic $In_{0.52}Al_{0.48}As/In_{0.7}Ga_{0.3}As$ HEMTs Under Cryogenic Conditions," *IEEE Electron Device Lett.* 30 (2009) 1024–1026.

R.S. Kumar, R. Poornachandran, S. Baskaran, N.M. Kumar, S. Sandhiya, K.U. Shanmugapriya, "DC and RF Characterization of InAs based Double Delta Doped MOSHEMT Device," In *2018 IEEE Electron Devices Kolkata Conf.* (2018) 509–512.

C.I. Kuo, H.T. Hsu, C.Y. Wu, E.Y. Chang, Y. Miyamoto, Y. Chen, D. Biswas, "A 40-nm-Gate InAs/In0.7Ga0.3As Composite-Channel HEMT with 2200 mS/mm and 500-GHz fT," In *2009 IEEE Int. Conf. Indium Phosphide Relat. Mater.* (2009) 128–131.

Q. Li, X. Zhou, C.W. Tang, K.M. Lau, "High-Performance Inverted In0.53Ga0.47As MOSHEMTs on a GaAs Substrate With Regrown Source/Drain by MOCVD," *IEEE Electron Device Lett.* 33 (2012) 1246–1248.

P.S. Park, S. Rajan, "Simulation of Short-Channel Effects in N- and Ga polar AlGaN/GaN HEMTs," *IEEE Transactions on Electronic Devices* 58, no. 3 (2011) 704–708.

S.K. Pati, H. Paradeshi, G. Raj, N. M. Kumar, C.K. Sarkar, "Impact of Gate Length and Barrier Thickness on Performance of InP/InGaAs Based Double Gate Metal Oxide Semiconductor Heterostructure Field Effect Transistor (DG MOS-HFET)," *Superlattices and Microstructures* 55, no. 1 (2013) 8–15.

Y. Royter, K. Elliott, P. Deelman, R. Rajavel, D. Chow, I. Milosavljevic, C. Fields "High Frequency InAs-Channel HEMTs for Low Power ICs," *IEEE International Electron Devices Meeting 2003*, Washington, DC, USA (2003) 30.7.1– 30.7.4.

T. Takahashi, Y. Kawano, K. Makiyama, S. Shiba, M. Sato, Y. Nakasha, N. Hara, "Enhancement of f_{max} to 910 GHz by Adopting Asymmetric Gate Recess and Double-Side-Doped Structure in 75-nm-Gate InAlAs/InGaAs HEMTs," *IEEE Trans. Electron Devices* 64 (2017) 89–95.

G. Yu, Y. Cai, Y. Wang, Z. Dong, C. Zeng, D. Zhao, H. Qin, B. Zhang, "A Double-Gate AlGaN/GaN HEMT With Improved Dynamic Performance," *IEEE Electron Device Lett.* 34 (2013) 747–749.

9

Influence of Dual Channel and Drain-Side Recess Length in Double-Gate InAs HEMTs

Y. Vamsidhar

CONTENTS

9.1 Introduction

Low-power consumption devices are much preferable in high-frequency applications, such as satellite communication, owing to their superior efficiency. Recent research trends led to the fabrication of new novel devices based on III–V compound semiconductors emerging as a promising candidate because of their reduced noise figure at high-frequency radio frequency (RF) applications. Among the III–V compound semiconductors, high electron mobility transistors (HEMTs) show a great deal of potential facilitating superior performance such as higher transconductance, high current density, low power consumption, and higher operating frequency. Hence, this III–V HEMT can replace the conventional Si MOSFET in every application (Verma et al. 2019; Chao et al. 1990). InAs HEMTs show better performance like ultralow-power degradation and increased RF performance due to the high carrier mobility (Edward et al. 2008).

DOI: 10.1201/9781003093428-9

The direct current (DC) and RF performance of InAs-based single-gate HEMTs (SG-HEMTs) show some discrepancies like high gate leakage current, kink effect, and impact ionization. So a new approach is necessary to overcome these effects. Thus, double-gate (DG) architecture in the device surfaced as an excellent alternative to achieve higher performance (Ritesh et al. 2010; Vasallo et al. 2008). The short-channel effects are reduced by introducing this DG technique, providing high electrostatic control over the channel. It also reduces the output conductance (g_{ds}) because of the removed buffer in double-gate architecture.

The combination of the composite channel with reduced gate length in the architecture of DG-HEMTs exhibits a massive potential in sub-millimeter wave applications. The single-channel DG-HEMT device, combined with the thin barrier layer and a high indium concentration in the channel, can give outstanding frequency performance (Saravankumar et al. 2017). But this single-channel DG-HEMT experiences low-current levels in the device because of less sheet carrier density (n_s) in the channel. This can be mitigated by introducing multiple heterojunctions in the DG-HEMT to elevate the device's current driving capability and switching speed. Thus a dual channel is formed in the DG-HEMT, improving the carrier density in the channel. This peak sheet carrier density enables the device to achieve a high drain current and transconductance. The peak transconductance observed will significantly improve the device's RF performance (Chugh et al. 2017).

9.2 Structure of Dual Channel and Drain-Side Recess Length of Gate

A five-layered composite channel in the combination of $In_{0.7}Ga_{0.3}As/InAs/In_{0.7}Ga_{0.3}As/InAs/In_{0.7}Ga_{0.3}As$ is introduced in the device architecture to form a dual channel in the device. The sheet carrier density is significantly improved due to the formation of two 2DEGs in the device channel. The influence of this dual channel in the device is observed through the device DC and RF performance.

In this device, the distance between the gate and source ($L_{sg} + L_{rs}$) is fixed at 350 nm, and the recess length of the gate at the source side (L_{rs}) is kept constant at 30 nm, whereas the drain-side recess length of the gate (L_{rd}) is varied from 40 to 60 nm to optimize L_{rd}. Thus by reducing the gate recess length on the drain side, the DC and RF performance of the device will be significantly improved with reduced ON-resistance (R_{ON}). The influence of a dual channel with varying L_{rd} to obtain enhanced DC and RF performance is analyzed.

9.3 Formation of Dual Channel

The sheet carrier density (n_s) in the upper channel of the dual-channel HEMT is given by

$$n_{s_{1(x,y)}} = \frac{\varepsilon(x)}{qd_1}\left(V_{gs} - V_{th1}(x) - V_{c1}(y)\right) \tag{9.1}$$

where $V_{c1}(y)$ is the potential of the channel at any point along the channel, V_{th1} is the threshold voltage of the upper channel, V_{gs} is defined as the gate–source voltage, $\varepsilon(x)$ is defined as the permittivity of $In_xGa_{1-x}As$, $x = 0.7$ is the mole fraction of indium utilized, and d_1 is defined as the distance between the upper channel to the top gate of the device.

The sheet carrier density in the 2DEG of the lower channel is given by

$$n_{s_{1(x,y)}} = \frac{\varepsilon(x)}{qd_2}\left(V_{gs} - V_{th2}(x) - V_{c2}(y)\right) \tag{9.2}$$

where d_2 is the distance between the lower channel to the bottom gate of the device, V_{th2} is defined as the lower channel threshold voltage, and $V_{c2}(y)$ is the potential of the lower channel.

9.4 Performance of Double-Gate HEMTs

9.4.1 Sheet Carrier Density

The effect of varying the drain-side recess length of the gate on the conduction band offset (CBO) performance of a DCDG-HEMT at $L_g = 30$ nm is displayed in Figure 9.1. This figure visualizes the formation of a dual channel in the device. Compared with single-channel HEMT, the dual-channel carriers are very high, leading to peak sheet carrier density (n_s). The increase in the triangular quantum depth is observed for high gate drain-side recess length, resulting in a significant reduction of sheet carrier density leading to reduced drain current.

9.4.2 DC Performance of DCDG-HEMTs

9.4.2.1 Transfer Characteristics

The transfer characteristics (I_{ds}–V_{gs}) of DCDG-HEMTs for varying drain-side recess lengths of 40–60 nm at $V_{gs} = -0.5$ to 1 V and constant $V_{ds} = 0.5$ V are visualized in Figure 9.2. For varying L_{rd} of 40, 50, and 60 nm, the respective transconductance (g_m) observed are 4.77, 4.66, and 4.53 S/mm. The increase

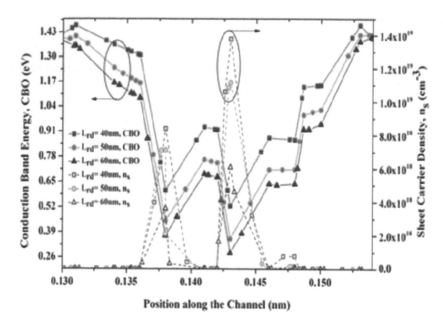

FIGURE 9.1
The conductance band energy is plotted as a function of position along the channel.

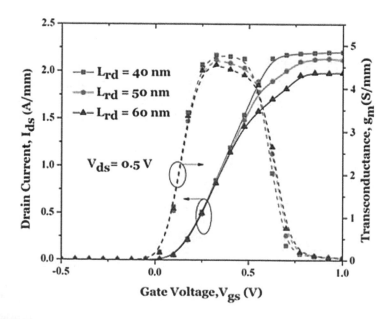

FIGURE 9.2
The transfer characteristics of dual-channel HEMTs. A reduction in drain-side recess length increases the drain current as well as transconductance.

in the drain-side recess length of the gate increases the drain resistance (R_D), which subsequently decreases the drain current. The peak transconductance of 4.77 S/mm is observed and a drain current of 2.203 mA for the L_{rd} = 40 nm (Takahashi et al. 2017).

Figure 9.3 shows the comparison of transfer characteristics of various devices, that is, SCSG-HEMT, SCDG-HEMT, and DCDG-HEMT for L_{rd} = 40 nm at V_{gs} = –1 to 1 V and V_{ds} = 0.5 V. Owing to high sheet carrier density, the DCDG-HEMT exhibits better transconductance and drain current compared with the SCSG-HEMT and SCDG-HEMT counterparts.

9.4.2.2 Threshold Voltage and Transconductance

The fluctuation of the threshold for different values of L_{rd} at V_{ds} = 0.5 V for the SCDG-HEMT and DCDG-HEMT is shown in Figure 9.4. The threshold voltage (V_T) of the DG-HEMT is explained by

$$V_{TH} = \left(\phi_B - \Delta E_C - \frac{N_\delta d_d^2}{2\varepsilon_d} \right) \tag{9.3}$$

where V_{TH} is defined as the threshold voltage, the height of the Schottky barrier is given by ϕ_B, E_C gives the discontinuity in the conduction band, N_δ is

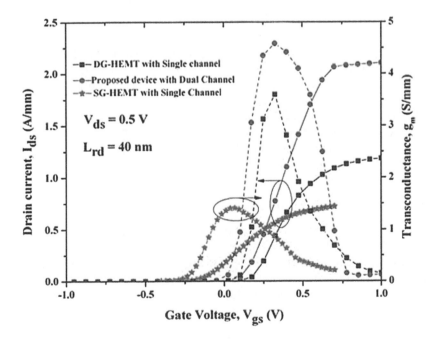

FIGURE 9.3

The variation in transfer characteristics of the single-channel and the dual-channel DG-HEMT. It shows that the DCDG-HEMT has a better response than the SCSG-HEMT providing improved transconductance and drain current.

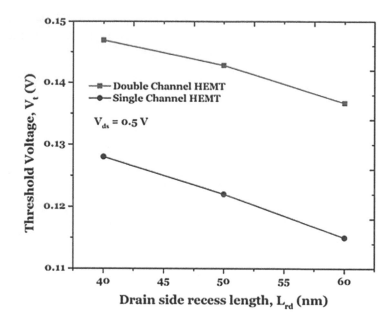

FIGURE 9.4
The threshold voltage is higher in dual-channel DG-HEMTs than single-channel DG-HEMTs. The figure depicts the threshold variation with drain-side recess length.

the density of the silicon doping layers, d_d is defined as the distance between the channel and gate of the device, and the permittivity of the barrier layer is given by ε_d. The threshold voltage of DCDG-HEMT is higher than SCDG-HEMT and starts decreasing for increased L_{rd}.

The comparison of threshold voltage and transconductance of the DCDG-HEMT with SCDG-HEMT is displayed in Figure 9.5. Because of the high electron density in the double well and high electron mobility of InAs, the DCDG-HEMT has the highest g_m and V_{TH} values.

9.4.2.3 Output Characteristics

The output characteristics (I_{ds}–V_{ds}) of the DCDG-HEMT for L_g = 30 nm and constant V_{gs} = 0.2, 0.4, and 0.6 V with varying L_{rd} are displayed in Figure 9.6. The saturated drain currents of 1.653, 1.628, and 1.605 A/mm are obtained at higher V_{gs} of 0.6 V for L_{rd} = 40, 50, and 60 nm, respectively. As L_{rd} increases, the drain current (I_{ds}) decreases due to poor electrostatic gate control over the channel. On the other hand, this enables InAs DCDG-HEMTs to operate efficiently at low drain bias resulting in reduced DC power dissipation, and hence they are very suitable for low-power applications (Matsuzaki et al. 2007).

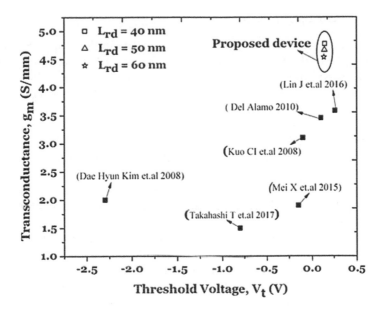

FIGURE 9.5
The DCDG-HEMT has a higher threshold voltage compared to the SCDG-HEMT. The figure shows the change of transconductance with the shift in threshold voltage of different devices compared with our proposed device.

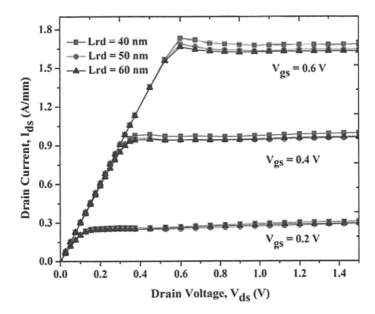

FIGURE 9.6
The output characteristics of (I_{ds}–V_{ds}) of the DCDG-HEMT. With the increase of V_{gs}, I_{ds} increased significantly with low L_{rd}.

9.4.2.4 Short-Channel Effects

Figure 9.7 displays the comparison of subthreshold slope (SS) of the DCDG-HEMT with SCDG-HEMT for L_g = 30 nm at V_{ds} = 0.5V for varying L_{rd}. The figure shows that the SS increases from 65 to 71 mV/decade for an increase in L_{rd}. A low L_{rd} also increases the electrostatic control over the channel, which helps in reducing the leakage current at the OFF state condition. For the same physical parameters, due to enhanced electrostatic control, the DCDG-HEMT yields an SS of 65 mV/decade for L_{rd} of 40 nm, whereas the SCDG-HEMT exhibits an SS of 68 mV/decade for the same L_{rd}.

The drain-induced barrier lowering (DIBL) is analyzed for the SCSG-HEMT and DCDG-HEMT with the same device physical parameters, as displayed in Figure 9.8.

For the optimized drain-side recess length L_{rd} of 40 nm, the DCDG-HEMT achieved a DIBL of 68 mV/V. But for the standard SCDG-HEMT, a DIBL of 71 mV/V is observed; the better DIBL in the DCDG-HEMT is due to the high ΔV_T achieved. The reduced short-channel effects (SCEs) in the dual-channel DG-HEMT show great potential compared to the SC counterparts. The better confinement of electrons in the channel by influencing the bottom heterojunction barrier improves the SCEs.

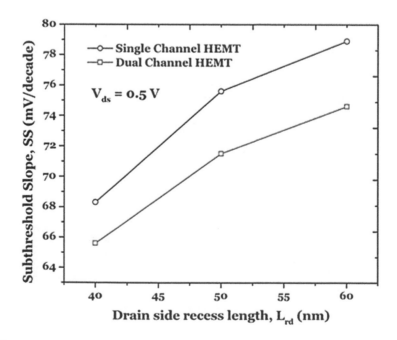

FIGURE 9.7
The subthreshold slope of the DCDG-HEMT and SCDG-HEMT for varying L_{rd}. The figure shows that the SCDG-HEMT has a higher subthreshold slope compared to the DCDG-HEMT.

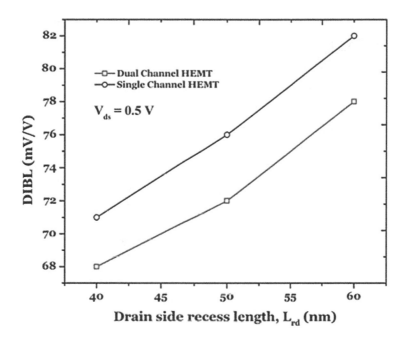

FIGURE 9.8
Comparison of drain-induced barrier lowering. The single-channel DG-HEMT has high drain-induced barrier lowering compared to the dual-channel DG-HEMT.

9.4.3 RF Performance of DCDG-HEMTs

The frequency performance of the DCDG-HEMT is characterized by using the maximum frequency of oscillation (f_{max}), transition frequency (f_T), and the intrinsic delay due to the presence of RF parasitics at higher frequencies (Mohankumar et al. 2010). The values of output conductance (g_{ds}) and transconductance play a crucial role in determining the aforementioned performance metrics.

The variation of the f_{max} with V_{gs} for different L_{rd} at $V_{ds} = 0.5$ V is displayed in Figure 9.9. For $L_{rd} = 60$ nm, f_{max} achieves peak value with reduced parasitics. The g_{ds} is proportional to the f_{max}. The variation of g_{ds} for the varying drain-side recess length is given as a bar diagram in Figure 9.10. The figure shows that for high L_{rd}, the g_{ds} of the device is reduced. This low g_{ds} helps in achieving a high maximum frequency of oscillation. The DCDG-HEMT is compared with the conventional SCDG-HEMT for the same physical parameters at an optimized L_{rd} of 60 nm in Figure 9.11. The figure shows that the DCDG-HEMT provides better f_{max} compared to SCDG-HEMT due to reduced parasitic and SCEs.

The variation of cutoff frequency (f_T) with V_{gs} with $L_g = 30$ nm and $V_{ds} = 0.5$ V for the varying L_{rd} is displayed in Figure 9.12. In general, transconductance is directly proportional to the f_T. Because of the high sheet carrier

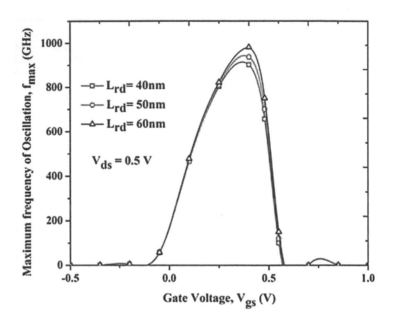

FIGURE 9.9
The variation of maximum frequency of oscillation (f_{max}) of the DCDG-HEMT with V_{gs} for different L_{rd}. f_{max} achieves its peak at L_{rd} = 60 nm.

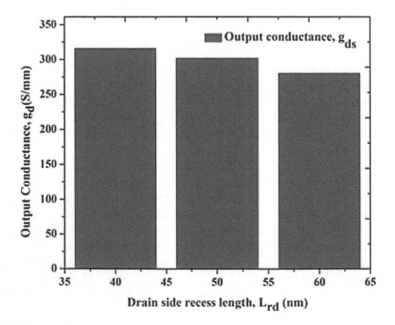

FIGURE 9.10
The different output conductance with different drain-side recess lengths. It is more for the lower drain-side recess lengths in dual-channel HEMTs.

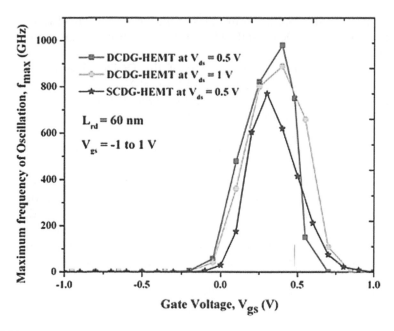

FIGURE 9.11
The variation of f_{max} with V_{gs} and $L_{rd} = 60$ nm for SCDG-HEMTs and DCDG-HEMTs. The DCDG-HEMT has a better f_{max} compared to the SCDG-HEMT.

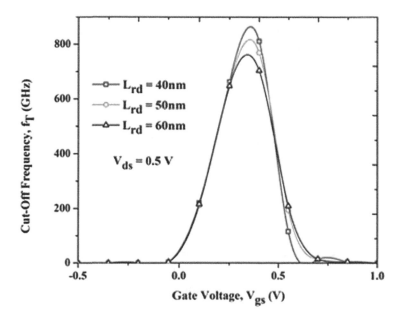

FIGURE 9.12
The variation of cutoff frequency (f_T) with V_{gs} with $L_g = 30$ nm and $V_{ds} = 0.5$ V for varying L_{rd}. The cutoff frequency decreases with increased drain-side recess length for the DCDG-HEMT.

density for L_{rd} = 40 nm, high transconductance is noted. This high transconductance enables the DCDG-HEMT with L_{rd} of 40 nm to acquire peak cutoff frequency.

The comparison of cutoff frequency of the DCDG-HEMT and SCDG-HEMT with V_{gs} and optimized L_{rd} of 40 nm at V_{ds} = 0.5 V is displayed in Figure 9.13. The comparison is also made for the dual-channel device at V_{ds} of 1 V. The comparison clearly shows the potential of DCDG-HEMT with superior RF performance than the SCSG-HEMT. This outstanding frequency of DCDG-HEMT is mainly due to the reduction in short-channel effects and reduced parasitic capacitances.

The variation of f_{max} and f_T for varying gate lengths at L_{rd} = 40 nm and V_{ds} = 0.5 V is shown in Figure 9.14. This figure shows that for a reduced gate length of 30 nm, peak f_T = 810 GHz and f_{max} = 900 GHz is observed. Because of the excellent gate controllability demonstrated by the double-gate architecture, InAs-based DCDG-HEMTs show great potential in high frequency and empower them as good candidates for future wireless communication systems.

The performance parameters for the SCDG-HEMT and DCDG-HEMT are illustrated as a comparison table in Table 9.1.

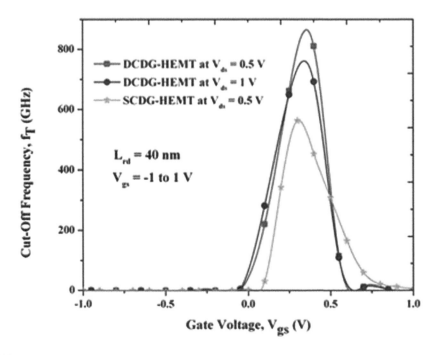

FIGURE 9.13
The comparison of cutoff frequency of DCDG-HEMTs and SCDG-HEMTs at an optimized L_{rd} of 40 nm at V_{ds} = 0.5 V.

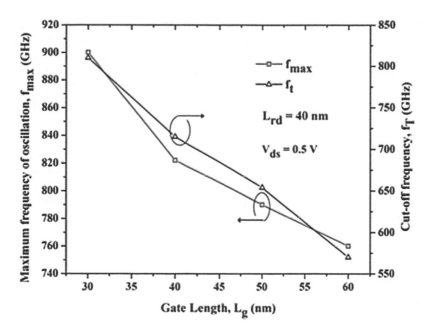

FIGURE 9.14
The variation of f_{max} and f_T for varying gate lengths at $L_{rd} = 40$ nm and $V_{ds} = 0.5$ V. This figure shows that for a reduced gate length of 30 nm, a peak $f_T = 810$ GHz and $f_{max} = 900$ GHz are observed.

TABLE 9.1

Comparison of Various Performance Parameters Analyzed for the SCDG-HEMT and DCDG-HEMT

Parameter	SCDG-HEMT	DCDG-HEMT
g_m	1.4 mS/μm	4.77 ms/μm
V_T	0.12 V	0.1469 V
I_{ON}	1.4 mA/μm	2.203 mA/μm
SS	68 mV/decade	65 mV/decade
DIBL	71 mV/V	68 mV/V
f_{max}	860 GHz	900 GHz
f_T	750 GHz	810 GHz

9.5 Summary

The influence of the DCDG-HEMT for varying drain-side recess lengths (L_{rd}) is analyzed using the technology computer-aided design (TCAD) simulator tool. The performance of the conventional DG-HEMT is enhanced by introducing the five-layered channel $In_{0.7}Ga_{0.3}As/InAs/In_{0.7}Ga_{0.3}As/InAs/In_{0.7}Ga_{0.3}As$ that forms a dual channel in the device. Due to the peak value of n_s observed, this device delivers a high drain current and g_m, with reduced SCEs. The InAs-based DCDG-HEMT device with gate length of 30 nm facilitated improved performance in terms of V_T = 0.1469V, g_m = 4.77 S/mm, I_{ON} = 2.203mA/µm, DIBL = 68 mV/V, and SS = 65 mv/decade. The RF performance metrics of the DCDG-HEMT are f_{max} = 900 GHz and f_T = 810 GHz, which are relatively high compared to other devices.

The optimization of L_{rd} greatly improves the f_{max} by increasing the output conductance to achieve a better drain current due to the reduced drain resistance (R_D). The DC and RF performance of the DCDG-HEMT adheres with great potential for low-power applications, terahertz frequency, high-speed applications, and satellite communications systems.

References

Chao PC, Tessmer AJ, Duh KHG, 1990, "W-band low-noise InAlAs/InGaAs," *IEEE Electron Device Lett.*11(1), 59–62.

Chugh N, Bhattacharya M, Kumar M, Gupta RS, 2017, "Sheet carrier concentration and threshold voltage modeling of asymmetrically doped AlGaN/GaN/AlGaN double-heterostructure HEMT," 4th IEEE Uttar Pradesh Section International Conference Electronics Computational Electronics, 446–451.

Edward YC, Ching IK, Hwei PH, Chia-Yuan C, 2008, "InAs/In1-xGaxAs composite channel high electron mobility transistors for high speed applications," *European Microwave Integrated Circuit Conference*, 198–201.

Matsuzaki H, Maruyama T, Koasugi T, Takahashi H, Tokumitsu M, Enoki T, 2007, "Lateral scale down of InGaAs/InAs composite-channel HEMTs with tungsten-based tiered ohmic structure for 2-S/mm gm and 500-GHz f_T," *IEEE Transactions on Electron Devices*, 54, 378–384.

Mohankumar N, Syamal B, Sarkar CK, 2010, "Influence of channel and gate engineering on the analog and RF performance of DG MOSFETs," *IEEE Transactions on Electron Devices*, 57, 820–826.

Ritesh G, Rathi S, Gupta M, Gupta RS, 2010, "Microwave performance enhancement in double and single gate HEMT with channel thickness variation," *Superlattices and Microstructures*, 47(6), 779–794.

Saravana Kumar R, Mohanbabu A, Mohankumar N, Godwin Raj D, 2017, "Simulation of InGaAs subchannel DG-HEMTs for analogue/RF applications," *International Journal of Electronics*, 105(3), 446–456.

Takahashi T, Kawano, Yoichi, Makiyama, Kozo, Shiba, Shoichi, Sato, Masaru, Nakasha, Yasuhiro, Hara, Naoki, 2017, "Maximum frequency of oscillation of 1.3 THz obtained by using an extended drain-side recess structure in 75-nm-gate InAlAs/InGaAs high-electron-mobility transistors," *Appl. Phys. Express*, 10, 0–4.

Vasallo BG, Wichmann N, Bollaert S, 2007, "Comparison between the dynamic performance of double and single-gate AlInAs/InGaAs HEMTs," *IEEE Transactions on Electron Devices*, 54(11), 2815–2822.

Verma YK, Mishra V, Verma PK, Gupta SK, 2019, "Analytical modeling and electrical characterization of ZnO based HEMTs," *International Journal of Electronics*, 106, 707–720.

10

Noise Analysis in Dual Quantum Well InAs-based Double-Gate (DG) HEMTs

Girish Shankar Mishra

CONTENTS

10.1 Introduction

The high-speed performance of high electron mobility transistors (HEMTs) is obtained by utilizing material with good physical properties that improve the sheet carrier density in the two-dimensional electron gas (2DEG) and increase electron velocity in the device (Mateos et al. 2015). The high electron mobility characteristics of indium arsenide(InAs) (40,000 cm^2/V-s) make it a promising candidate for the current semiconductor device research trends for future wireless communications. These outstanding advantages of InAs emerged as a suitable material for the HEMT architecture, and such devices are being used in low-noise, low-power, and high-frequency applications (Chiu et al. 2015; Tsai et al. 2003).

Conventional single quantum well DG-HEMTs with short gate length structure and a composite channel having a higher concentration of indium exhibits excellent radio frequency (RF) performance metrics. The thin InAlAs barrier and narrow bandgap of the indium material achieve superior values of maximum frequency of oscillation (f_{max}) and transition frequency (f_T) (Saravanakumar et al. 2017). Low current levels are observed in the single quantum well HEMT with limited sheet carrier density concentration in the 2DEG.

The current driving capability of the single quantum well HEMT is improved by incorporating multiple heterostructure techniques in the

DOI: 10.1201/9781003093428-10

device. These multiple heterostructures in the device enhance the device performance by creating a dual quantum well in the channel. This dual quantum well increases the sheet carrier density (n_s) to achieve high transconductance (g_m) and drain current, which in turn help the device to achieve better RF performance (Chugh et al. 2017).

10.2 Dual-Quantum-Well Structure

A five-layered composite channel in the combination of $In_{0.7}Ga_{0.3}As/InAs/In_{0.7}Ga_{0.3}As/InAs/In_{0.7}Ga_{0.3}As$ is introduced in the proposed architecture to form a dual quantum well in the device, which enables the formation of two 2DEGs in the device channel, which significantly improves the device sheet carrier density. The influence of this dual quantum well in the device is observed through the device's noise performance.

10.3 Green's Function Method

Green's function method from the technology computer-aided design (TCAD) simulator tool efficiently analyzes the proposed device noise performance under large-signal conditions. The flicker noise (1/f) and generation and recombination noise (G-R noise) are the two predominant noise sources in the HEMTs. The device G-R noise is due to channel fluctuations (Kugler 1998). The defect scattering in the device channel can vary the carrier mobility, which acts as a source of flicker noise in the device (Fleetwood 2015).

The spectral density of the variation in the resistance and its relation with the G-R noise is given by

$$\frac{S_R(\omega)}{R^2} = \frac{S_G(\omega)}{G^2} = \frac{S_N(\omega)}{N_0^2} \cdot \frac{\overline{\delta N^2}}{N_0} \cdot \frac{4\,T_N}{1+(\omega T_N)^2} \tag{10.1}$$

where $S_R(\omega)$ is the spectral density of the resistance, $S_N(\omega)$ is the spectral density of the number of carriers, $S_G(\omega)$ is the spectral density of the conductance, ω is the circular frequency of the device, and δ is the variation in the number of carriers. In contrast, T_N is the carrier's lifetime. Another critical design attribute is the total noise figure, expressed as

$$\text{Noise figure}(NF) = 10\log_{10}\left(\frac{S_i/N_i}{S_0/N_0}\right) \tag{10.2}$$

where S_i is the signal at the input, N_i is the noise at the input, S_o is the signal at the output, and N_o is the noise at the output.

The InAs DWDG-HEMT with a gate length of 30 nm exhibits a high g_m of 4.77 S/mm, f_T of 810 GHz, and a f_{max} of 900 GHz, and a reduced noise performance of $NF_{min} = 1.62$ dB at 810 GHz for $V_{gs} = 0.6$ V and $V_{ds} = 0.5$ V. These improved noise characteristics make them suitable for low-noise amplifiers (LNAs) for high-frequency applications.

10.4 Noise Analysis

The noise models included in the alternating current (AC) modulation technique available in the physics section of SDEVICE are used in this work to simulate the noise in the proposed device. The AC technique is included to analyze the noise over a wide range of frequencies. The starting frequency is 0.1 Hz, and the ending frequency is 10^{13} Hz. From this simulated data, the various noise parameters are extracted. The Poisson solver, coupled to the HD model, is used to calculate the transistor's dynamic response, and the spectral range of the simulation is limited to frequencies in the sub-terahertz (THz) range.

The spectral densities of the voltage and current fluctuations are analyzed along with the small-signal characteristics to study the proposed device noise behavior. The gate and drain bias is varied to compute the gate voltage spectral density (S_{vg}), drain voltage spectral density (S_{vd}), gate current spectral density (S_{ig}), and drain current spectral density (S_{id}) over a wide range of frequencies. A useful noise characterization tool is obtained through this measurement technique, and this simulated data reports the defects and evolution of the HEMTs under investigation (Wirth et al. 2005).

The variation of gate voltage spectral density (S_{vg}) and drain voltage spectral density (S_{vd}) of DWDG-HEMTs as a function of the frequency at $V_{ds} = 0.5$ V and $V_{gs} = 0, 0.3, 0.6, 0.9$, and 1.2 V is shown in Figure 10.1. At the gate-insulator/barrier and channel/barrier interface, the traps are found abundantly, leading to increased scattering of carriers in the channel, subsequently decreasing the drain current.

The drain current (I_{ds}) degradation is due to the fluctuation of voltages caused by the interface traps. The S_{vg} and S_{vd} are used to characterize these voltage fluctuations in the gate and drain of the device. The maximum values of S_{vg} and S_{vd} are obtained at $V_{gs} = 0.6$ V at lower frequency regions. In general, the spectral densities of the noise are inversely proportional to the frequency, i.e., $S \propto 1/f$ (Liu et al. 2014). Because of this proportionality, S_{vg} and S_{vd} decrease at the higher frequency regions.

The variation of gate current spectral density (S_{ig}) and drain current spectral density (S_{id}) with frequency for varying $V_{gs} = 0, 0.3, 0.6, 0.9$, and 1.2 V, and

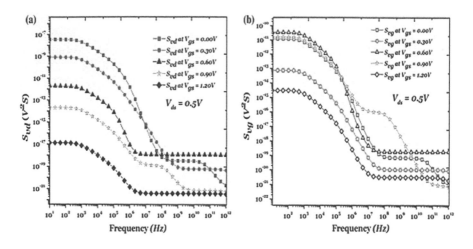

FIGURE 10.1

(a) The drain–voltage spectral density with the change of frequency variation. It is high when the gate–source voltage is low. (b) The gate-voltage spectral density interpretation for the double-gate HEMT has been depicted. The drain–source voltage is kept constant at 0.5 V.

V_{ds} = 0.5 V is shown in Figure 10.2. The number of G-R centers at the interface of InGaAs/InAs/InGaAs/InAs/InGaAs is relatively high. The varying gate bias on the conduction band explains the characteristics of the S_{id} of the proposed device. The maximum S_{id} is observed at the gate bias, where g_m achieved its peak. When V_{gs} approaches the pinch-off condition, the maximum S_{id} is observed, and the electrons in the conduction band are bound to the interface. The electrons are emitted from the active trap centers to the conduction band due to the device G-R noise. Active trap centers then capture the emission of a hole from the valence band to the empty trap center (G) at the valence band (R) (Saha 2015). This phenomenon provokes the drain current variation due to the introduction of charge carriers in the channel, increasing the S_{id}.

At high voltages, electrons will enter randomly into the device because of the influence of high leakage current, which acts as a source of shot noise in the device. It is evident from Figure 10.2 that S_{id} is marginally greater than S_{ig} at higher values of V_{ds}. The carrier thermal noise arises due to the fluctuation of carriers in the channel. The increment in the carrier fluctuation is due to the effect of hot electrons generated with an increase in V_{gs}. The presence of these hot carriers is the reason for increment in both the S_{id} and S_{ig}, and is found to be in good agreement with the mechanism proposed in.

A linear two-port noiseless circuit is introduced to characterize the noise sources in the proposed device as shown in Figure 10.3 (Hsu et al. 2002). By using this two-port network, the various noise parameters in the DWDG-HEMT are explored. From this analysis, three crucial noise parameters – the noise conductance (G_n), the minimum noise figure (NF_{min}), and the optimum

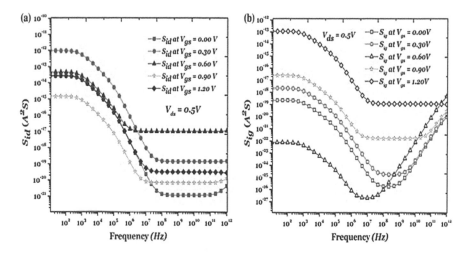

FIGURE 10.2
The variation of (a) drain current spectral density and (b) gate current spectral density of a DWDG-HEMT with frequency at a constant drain–source voltage (V_{ds}) = 0.5V and gate–source voltage (V_{gs}) = 0, 0.3, 0.6, 0.9, and 1.2 V.

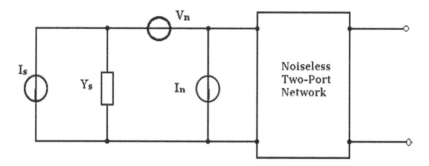

FIGURE 10.3
Two-port noise equivalent circuit with noise sources (V_n, I_n) at the input.

source admittance (Y_{opt}) – are evaluated. The total noise figure of the proposed device is obtained by using Equation 10.3

$$NF = NF_{min} + \frac{R_n}{G_S}\left(G_S - G_{opt}\right)^2 + \frac{R_n}{G_S}\left(B_S - B_{opt}\right)^2 \tag{10.3}$$

The Y_{opt} is the optimum source admittance of the device expressed as Y_{opt} = $G_{opt} + jB_{opt}$, where G_{opt} is the optimum source conductance and B_{opt} is the optimum source susceptance (Zhu et al. 2008). By optimizing the G_S and B_S of the device, the minimum noise figure (NF_{min}) is obtained from Equation 10.3. The carrier concentration in the channel is increased for high V_{gs} and thus increases G_{opt} and B_{opt}. At the lower frequency region, both the G_{opt} and

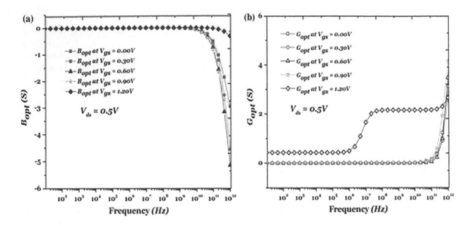

FIGURE 10.4

The optimum source susceptance and optimum source conductance versus the frequency of the DWDG-HEMT at V_{gs} = 0 to 1.2 V and V_{ds} = 0.5 V.

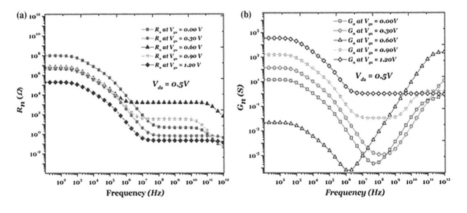

FIGURE 10.5

(a) The equivalent noise resistance variation of DWDG-HEMT with frequency and (b) the variation of equivalent noise conductance with frequency. Here drain–source voltage (V_{ds}) = 0.5 V, and the gate–source voltage (V_{gs}) varies from 0 V to 1.2 V.

B_{opt} are saturated. Still, Gopt and Bopt tend to decrease at the higher frequencies because of parasitics in the device, as depicted in Figure 10.4.

Figure 10.5 shows the variation of the equivalent noise resistance (R_n) and conductance (G_n) of the DWDG-HEMT with frequency for V_{gs} = 0 to 1.2 V and V_{ds} = 0.5 V. The decrease of R_n at the lower frequency region is due to high gate leakage current due to shot noise in the device. A flat response of R_n is observed for this DWDG-HEMT around the frequency region of 810 GHz. For broadband LNA design, the optimized value of R_n is very desirable to achieve good noise performance when the source impedance is mismatched to the optimum termination.

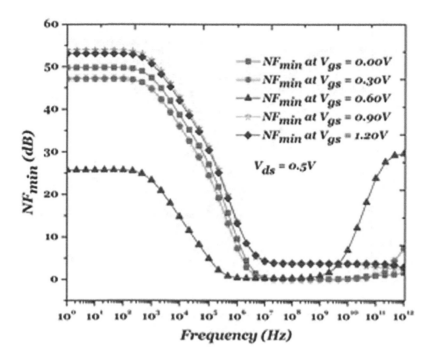

FIGURE 10.6
The variation of minimum noise figure (NF$_{min}$) of the InAs-based DWDG-HEMT with frequency for varying gate bias V$_{gs}$ = 0 to 1.2 V and constant V$_{ds}$ = 0.5 V.

The variation of minimum noise figure (NF$_{min}$) of the InAs-based DWDG-HEMT with frequency for varying gate bias V$_{gs}$ = 0 to 1.2 V and constant V$_{ds}$ = 0.5 V is shown in Figure 10.6. An increase in NF$_{min}$ observed at the lower frequency is mainly due to the effect of shot noise at the gate. The concentration of carriers in the dual quantum well increases with an increase in V$_{gs}$, thereby increasing the channel conductance. It efficiently reduces the noise at higher frequencies. An excellent flat response of NF$_{min}$ is achieved over a wide range of frequencies due to the availability of large bandwidth in the proposed device. The optimized NF$_{min}$ value of 1.62 dB at 810 GHz for V$_{gs}$ = 0.6 V and V$_{ds}$ = 0.5 V compared to single-quantum-well devices.

Figure 10.7 shows the overall noise figure (NF) as a function of the frequency of DWDG-HEMT for varying V$_{gs}$ = 0 to 1.2 V and V$_{ds}$ = 0.5 V. The NF and NF$_{min}$ are directly proportional as expressed in Equation 10.3. At the lower frequency region, the increase in NF is due to the rise in NF$_{min}$, and it is visible in Figure 10.7 (Mohanbabu et al. 2017). Comparison of the total noise figure of a single-quantum-well device with the proposed dual-quantum-well device for the optimized gate bias V$_{gs}$ = 0.6 and V$_{ds}$ = 0.5V is shown in Figure 10.8. Better noise performance is observed in dual-quantum-well devices due to increased transconductance and reduced parasitics.

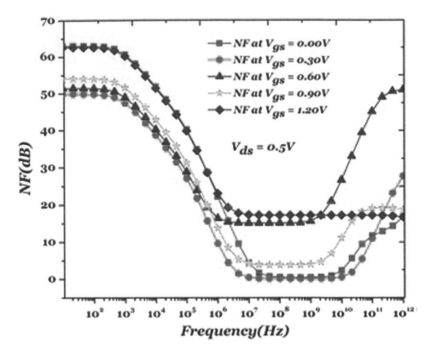

FIGURE 10.7
The overall noise figure (NF) as a function of the frequency of DWDG-HEMT for varying V_{gs} = 0 to 1.2 V and V_{ds} = 0.5 V.

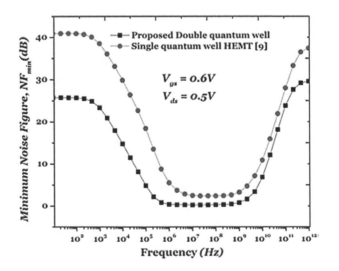

FIGURE 10.8
The comparison of the total noise figure of a single-quantum-well device with the proposed dual-quantum-well device for the optimized gate bias V_{gs} = 0.6 V and V_{ds} = 0.5 V.

10.5 Summary

In this chapter, the noise performance of dual quantum well InAs-based HEMT architecture is discussed. The influence of this dual quantum well for improved noise characterization is studied. The noise models included in the alternating current (AC) modulation technique available in the physics section of SDEVICE are used to simulate the noise analysis in the proposed device. The DWDG-HEMT exhibits better noise performance compared to the SWDG-HEMT, increasing the feasibility for high-frequency applications.

References

H.-C. Chiu, W.-Y. Lin, C.-Y. Chou, S.-H. Yang, K.-D. Mai, P. Chiu, W.J. Hsueh, J.-I. Chyi, "Device stress evaluation of InAs/AlSb HEMT on a silicon substrate with refractory iridium Schottky gate metal," *Microelectronic Engineering* 138 (2015) 17–20.

N. Chugh, M. Bhattacharya, M. Kumar, R.S. Gupta, "Sheet carrier concentration and threshold voltage modeling of asymmetrically doped AlGaN/GaN/AlGaN double heterostructure HEMT," *2017 4th IEEE Uttar Pradesh Sect. Int. Conf. Electr. Comput. Electron.* (2017) 446–451.

D.M. Fleetwood, "1/f noise and defects in microelectronic materials and devices," *IEEE Transactions on Nuclear Science* 62, no. 4 (Aug. 2015) 1462–1486.

S.S.H. Hsu et al., "Characterization and analysis of gate and drain low-frequency noise in AlGaN/GaN HEMTs," *Proceedings IEEE Lester Eastman Conference on High-Performance Devices*, Newark, DE, USA (2002) 453–460, doi: 10.1109/LECHPD.2002.1146787.

S. Kugler, "Generation-recombination noise in the saturation regime of MOSFET structures," *IEEE Transactions on Electron Devices* 35, no. 5 (May 1988) 623–628.

Y. Liu, Y. Zhuang, "A gate current 1/f noise model for GaN/AlGaN HEMTs," *J. Semicond.* 35 (2014).

J. Mateos, H. Rodilla, B.G. Vasallo, T. González, "Monte Carlo modelling of noise in advanced III--V HEMTs," *Journal of Computational Electronics* 14 (2015) 72–86.

A. Mohanbabu. R. Saravana Kumar, N. Mohankumar, "Noise characterization of enhancement-mode AlGaN graded barrier MIS-HEMT devices," *Superlattices and Microstructures* 112 (2017) 604–618.

S.D. Nsele, J.G. Tartarin, L. Escotte, S. Piotrowicz, S. Delage, "InAlN/GaN HEMT technology for robust HF receivers: an overview of the HF and LF noise performances," *2015 Int. Conf. Noise Fluctuations, ICNF 2015.* (2015), pp. 1–4, doi: 10.1109/ICNF.2015.7288538.

S.K. Saha, *Compact Models for Integrated Circuit Design: Conventional Transistors and Beyond*, CRC Press, Taylor &Francis, Boca Raton, USA, August 2015, pp. 41e42 ch. 2, sec. 2.2.6.2.

R. Saravana Kumar, A. Mohanbabu, N. Mohankumar, D. Godwin Raj, "Simulation of InGaAs subchannel DG-HEMTs for analog/RF applications," *International Journal of Electronics*, 105, no. 3 (2017) 446–456.

R. Tsai et al., "Metamorphic AlSb/InAs HEMT for low-power, high-speed electronics," *25th Annual Technical Digest 2003. IEEE Gallium Arsenide Integrated Circuit (GaAs IC) Symposium, 2003*, San Diego, CA, USA (2003) 294–297, doi: 10.1109/GAAS.2003.1252415.

G. I. Wirth, Jeongwook Koh, R. da Silva, R. Thewes, R. Brederlow, "Modeling of statistical low-frequency noise of deep-submicrometer MOSFETs," *IEEE Transactions on Electron Devices* 52, no. 7 (July 2005) 1576–1588.

Y. Zhu, C. Wei, O. Klimashov, B. Li, C. Zhang, Y. Tkachenko, "Gate width dependence of noise parameters and scalable noise model for HEMTs," *2008 European Microwave Integrated Circuit Conference*, Amsterdam (2008) 298–301.

Index